WONDERS OF THE SKY

Observing Rainbows, Comets, Eclipses, the Stars, and Other Phenomena

FRED SCHAAF

Dover Publications, Inc., New York

Published in Canada by General Publishing Company, Ltd., 30 Lesmill Road, Don Mills, Toronto, Ontario.
Published in the United Kingdom by Constable and Company, Ltd., 10 Orange Street, London WC2H 7EG.

Wonders of the Sky: Observing Rainbows, Comets, Eclipses, the Stars, and Other Phenomena is a new work, first published by Dover Publications, Inc., in 1983.

Book Design by Paula Goldstein

Manufactured in the United States of America
Dover Publications, Inc., 180 Varick Street, New York, N.Y. 10014

Library of Congress Cataloging in Publication Data

Schaaf, Fred.
Wonders of the sky.

Bibliography: p.
1. Astronomy—Popular works. 2. Meteorological optics—Popular works. I. Title.
QB44.2.S33 1983 523 82-14772
ISBN 0-486-24402-4

Contents

Preface

For over six years I have been writing a weekly newspaper column about current and upcoming sights in the sky. Among the memorable letters from readers of the column was one from a woman who wrote to tell me about her granddaughter's first sight of a rainbow: it seemed that even weeks after the event the young girl's mind and speech were still filled with the wonder of that experience.

To young children, much of the world is like the rainbow: fresh and potent, a kind of miracle leaping out of creation into great beauty and shining on with an ultimate mystery all the stronger for its not being too much abstracted or too far filed away in the cabinets of classification. What children at first have, and most adults largely lose, is the sense of *wonder*. But is it inevitable that gaining more experience of the world should result in loss of wonder? Not always. If a person resists the tyranny of thinking that all his factual knowledge and classifications are justifiably fixed and final, and always keeps his eyes open, he can retain some of that native wonder and bring it to his experience of a world far more complex and rich than that of any child. He can actually go on seeing at least some new sights all his life and, more importantly, he can find those views helping him to behold even the familiar parts of his world with fresh vision.

This book is an invitation to a tour of new sights and fresh visions. It is a survey which attempts to visit most of the phenomena of light in the sky, starting with the rainbow that may be as close and transient as a passing neighborhood shower, and working all the way out through the many layers of the sky (a multitude of heavens) to the naked-eye glow of the almost eternally enduring Great Andromeda Galaxy—a wheel of several hundred billion stars so distant that even light requires over two million years to make the trip to us from there. All a person needs in order to take in real life the tour set forth by this book is his or her unaided eyes and that all-important step: the step outside, leading to actually walking beneath the heavens and becoming intimately involved with these sights of beauty which beckon us to endless adventure and wonder.

That first step is essential. Facts take on greater life and open fresh perspectives to a person who actually experiences the things to which the facts apply. In consideration of this, my discussions in this book are mostly descriptive, with an emphasis on observation of phenomena. I have also striven to illustrate some of these descriptions with accounts of what I personally have seen and felt when I witnessed representative or especially excellent displays of these phenomena. Such accounts may help to show more clearly how to find and study the sights. They also show that these phenomena do not occur alone, but rather in a rich sensory environment which usually includes other natural happenings that can blend to augment the beauty of the principal phenomenon.

A person also brings a unique state of mind—thoughts and feelings and particular knowledge—to each observation, and this has effect on the total experience. For scientific purposes, such effects may need to be considered in determining what kind of objectivity was achieved in a given observation. I have indeed stressed in this book that a fairly inexperienced naked-eye observer can sometimes make sightings which have important scientific value—for instance, in the observation of meteor showers and fireballs, bright comets, auroras, volcanic twilights, halos, certain variable stars (including bright novas!), and many other phenomena. Usually the enthusiasm and other feelings which one brings to an observation need not interfere with its scientific fidelity. And the most important aim in a book like this—intended primarily though not exclusively for nonscientists—is to convey along with the information the kind of enjoyment and wonder which one can experience in seeking and finding these sights.

The enjoyment and wonder in seeing a bright meteor sliding across the starry heavens, or the Northern Lights searching the sky restlessly with rays of colored radiance, or an exquisitely slender crescent moon just touching the silhouette of a line of intricately branched trees—these are sufficient reasons for going out to observe. Is it impractical to look for rainbows and other odd phenomena of the sky? Yes, in the strictest sense of "impractical." But what is strictly practical is only important insofar as it makes us free to enhance our appreciation of life and the universe we live in. The fact that the rainbow has no direct "practical" or functional purpose is not a defect, but its great merit: the rainbow exists for that higher purpose of stirring us to wonder.

I believe that almost all readers of this book will find described here at least a few phenomena which they have never even heard of, much less seen. Yet some form of nearly all the phenomena can be observed at least once in the course of a year—if one remembers when and where to look, and is alert. But perhaps the greatest and most rewarding accomplishment is to rediscover those strongest but commonest things whose edge of beauty has been dulled for us by our becoming habituated to them, or rather to our very small comprehension of them. Besides such things as biological and purely meteorological light sources, there are two major omissions from this book's nearly complete tour of naked-eye light phenomena in the sky: I have chosen to deal with the sun and the moon only in some of their more unusual guises (such as eclipses)—for there is far too much to say about them to fit into the course of a tour like this one. I hope, however, that what I do have to say here, and what you will see of them in their special appearances, will help you see afresh the power and the beauty which sun and moon manifest in even their most common hour.

In making acknowledgments, I wish to thank Pat Witt for many a wonderful trip to twilight, and Nora McGee for her very quotable comments on bright stars. I also wish to acknowledge, and request, any criticisms or notable observations, which readers can send to me at the address of the important Dark Skies for Comet Halley program, given at the end of Chapter Six ("Comets"). Special thanks go to two contributors, Chuck Fuller and Chris Duke, old friends from astronomy-club days.

I wish to thank all the photographers whose work appears

in this book. Especially important were the contributions of Robert Burnham, Jr., and Robert Greenler, and the very large and generous one of Dennis Milon. I have admired these men's work for years, and it was a pleasure to correspond with them in the course of assembling this book. I also extend my warm thanks to my friend of long correspondence, Guy Ottewell: not merely for the use of his star charts, which are adapted here, but for continuing clarity and inspiration. Another great debt I owe is to my very good friend Steve Albers, most of all for years of conversation and shared enjoyments. I have benefited from his knowledge and enthusiasm immeasurably. The only greater acknowledgment I could possibly make here is the one of gratitude for having been fortunate enough to see so many of these wonders of the sky.

As this work goes to press, I think about books which first helped to excite my own interest in watching the sky. Starting when I was young, my reading of them kept safely kindled a natural fire of interest and enjoyment which I know will burn on in me always. It is my hope that the present book will play such a role in the lives of some other people, both young and old—for it is never too early or too late to begin to look at the sky. In the end, it is to these people that this book must be dedicated, and to the rest of us who, though having looked and studied longer, are also only just awakening to the wonder that is with us all.

Cumberland, N.J. FRED SCHAAF
September 1982

I.
IN THE EARTH'S ATMOSPHERE

CHAPTER ONE

The Rainbow

Late on a summer afternoon at the end of a shower or thunderstorm, the sun will often burst through in the west while cool raindrops are cascading down. When this happens a look to the west will show the rain transformed into a shimmering curtain of individually glittering jewels. But as the rain passes and the dark clouds march off, face the east. There, if you are lucky, you will see the pledge of peace at storm's end which is as grand and awesome as a thunderstorm but also as lovely and delicate as anything we know. It is one of those extra touches of beauty which seem to have no function but which nature has in every secret corner, every unexpected place. Here the secret corner is the receding edge of black storm, for on it there is an arc of intense colors embroidered with needles of sunlight onto the fabric of falling rain. White sunlight shines through the million windows of the raindrops, but each drop is mirroring and splitting the light into a full spectrum of hues to form one sky-spanning arc of colors. We are seeing the rainbow.

Although the rainbow has been seen by almost everyone, even the most alert and experienced observers are fortunate to spot it more than a few times each year. The rainbow is the most famous special phenomenon of meteorological optics, but few people know enough about rainbows to take full advantage of their infrequent occurrences. Most of us realize that each rainbow is unique, but there are many beautiful aspects of rainbows and many other types of bows which are not popularly known at all. There certainly can be more than one rainbow at a time—two are often visible to a single observer and in reality the whole sky may then be full of them. Rainbows can be caused by the moon and by the reflected image of the sun in water; other bows can occur in dew on the ground and in

Fig. 1. *A brilliant rainbow appears to descend vertically during a thunderstorm in Slater, Iowa.*

fog, and these various bows differ from the common rainbow in several striking ways. There is no better place to start a journey of wonders the eye can see in the sky than on the road of the rainbow: the path is infinite (the rainbow is a circle), but the passing shower which helps form it is often at our very doorstep.

The arc of colors with which we are most familiar is the *primary rainbow*. It forms a full circle (normally cut into an arc by the horizon) around a spot directly opposite the sun

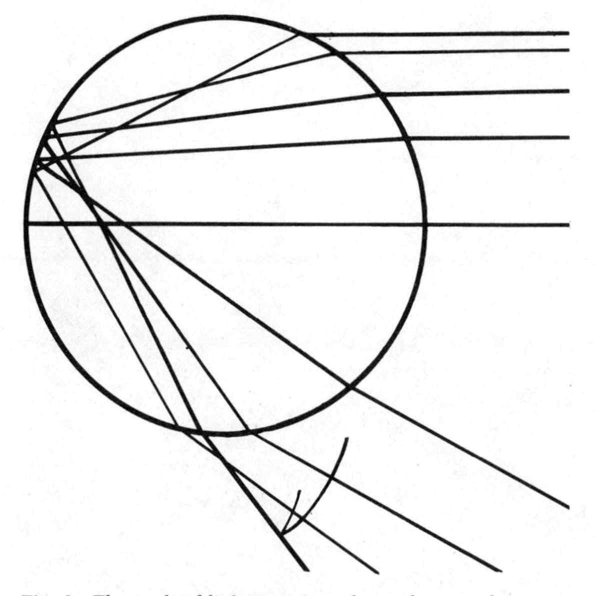

Fig. 2. *The path of light rays in a drop of water, forming a rainbow by concentration of emergent rays at the angle of minimum deviation. After Minnaert.*

(this *antisolar point* is located at the shadow of the observer's head). The primary rainbow always occurs at about 42° from this spot, and you can judge that distance by holding out your fist at arm's length and measuring just over four fist-widths (one fist-width equals about 10°, one-ninth of the distance from horizon to zenith). Only when the sun is less than 42° above the horizon can any part of the primary rainbow be in the sky—

therefore the bow is never visible around noon on a summer day in the United States or Europe. Of course, the situation is quite different from an airplane, for then there can easily be a sufficient number of illuminated raindrops below the observer and his horizon, and the full circle of the rainbow is sometimes visible.

The bow itself is a band of colors ranging from red on top to violet on bottom, but the claim that there are seven colors in the rainbow is not strictly true because there is really an infinite number of hues that merge into one another. The number of colors a person distinguishes depends partly on how intense and pure the particular bow is, but also on how one chooses to name various hues. Several hues which would be noticed and independently named by an artist and colorist might well be simply grouped together as *red* by most observers. The rainbow is formed when sunlight is refracted (bent) and reflected in raindrops. Since different wavelengths of visible light (colors) are refracted differently (violet is refracted more than red), the colors are spread out in a band slightly more than 1.5° wide, about three times the apparent diameter of the moon. It would, by the way, be quite possible to observe an ordinary sun-caused rainbow passing directly in front of the moon, though the sighting would be surpassingly rare. (A far less rare sight is a rainbow in what seems a cloudless, blue sky—in reality, of course, the cloud responsible for the rain causing the bow must be too thin to see or must have just dissipated.)

The beautiful primary bow is not alone. At a distance of about 51° from the antisolar point, there sometimes appears a second bow arching above the primary, but only one-tenth as intense (though it may seem relatively brighter or fainter than this depending on the background of clouds and other factors). The secondary bow forms from an extra reflection in raindrops and is fainter because much less light survives this longer and more absorbing trip and because the rays' more nearly tangential encounters with the droplet disperse them to form a wider band. The most notable other feature of the secondary bow besides its width—almost twice that of the primary—is the order of its colors—from violet on top to red on bottom, the opposite of the primary bow (thus the two reds are on the inside, facing each other). At least a trace of this second rainbow is often visible, but sometimes it can be beautifully prominent.

When the secondary bow is clearly visible, one is often struck by the fact that the sky between the bows is darker than

Fig. 3. *Given enough distance for the sun's rays to catch raindrops, rainbows can actually be seen below the horizon, sometimes forming a 360° circle.*

that outside. This is not an illusion. The two bows are the angle of minimum deviation for refracted light that is reflected once (primary bow) or twice (secondary) but some, though less, light emerges at greater angles. For the primary that light is inside the arc, for the secondary it is outside. And so the space between the bows is comparatively dark. This region is sometimes called *Alexander's dark band*, after Alexander of Aphrodisias, who noted it about eighteen hundred years ago.

If there are two bows caused by one and two reflections in raindrops respectively, could there be more? Yes! Rainbows of higher order are seen very rarely because even fainter than the secondary. Each higher-order bow is indeed wider still than the one before it. The third-order bow appears in the direction of the sun and has been observed a number of times, but the fourth-order bow has never been seen in the sky. If you ever should notice rainbow colors inside Alexander's dark band, you are probably seeing one of the rarest sights possible, a rainbow of the fifth order! A fifth-order rainbow is about 7° wide (its red overlaps with that of the secondary bow). Only one person, the nineteenth-century scientist Eleuthère Mas-

Fig. 4. *A prominent rainbow curves over the campus of Cornell University; at least one complete and one incomplete supernumerary arc run along just below the primary bow. Photographs by Chuck Fuller.*

cart, is recorded as having probably seen the fifth-order rainbow in nature. There are countless still higher orders but none of these has ever been glimpsed except by artificial means. A fine article in the July 1977 issue of *Scientific American* (Jearl Walker in "The Amateur Scientist") describes a fairly simple experiment in which a dozen rainbows can be seen in a single drop of water. In 1868 Felix Billet observed rainbows up to and including that of the 19th order in a thin stream of water produced by an apparatus he had made. All of these higher-order rainbows occur whenever the primary does; if their contrast with the background were only greater, we could sometimes see the whole sky fill with rainbows of various widths. Billet made a diagram showing the position and breadth of the first nineteen rainbows as they would appear around the sky or around a drop of water. He called his diagram a *rose of rainbows*.

The rainbows of higher order may be extremely rare in nature, but perhaps like so many natural phenomena they seem more infrequent than they really are because so few people know where to look for them, and because too many people are altogether neglectful of the natural beauty springing up and shining everywhere about them. But however uncommon the higher-order bows may or may not be, we can still often see more than the two first bows themselves: we can sometimes see the *supernumerary arcs*. These arcs are extra bands of color found on the inside of the primary bow and, quite rarely, on the outside of the secondary. There is sometimes a gap between the lowest color of the primary (violet) and the first supernumerary arc, but often the arcs appear as a continuation of the bow, and may therefore be overlooked by many people. The arcs, however, are alternately pink and green (sometimes the colors may be a little off or faded), and therefore a simple test for the presence of them is to note whether there is any pink or red near the bottom of a fully developed primary rainbow. If there is. it is surely a supernumerary arc, for the red of the primary bow is always found on the top. As many as five of these arcs have been seen with the primary bow, the number visible correlating with the size of the raindrops involved. According to M. Minnaert (pp. 178–179),* very large drops (1–2 mm in diameter) produce the most supernumerary arcs and also "very bright violet and vivid green" and "pure red, but scarcely any blue" in the primary bow. Even in these best displays, the arcs are seldom visible along the entire length of the bow. They are most often seen beneath the highest part of the primary's arch.

The supernumerary arcs actually should not be considered as separate bows or as additional bands of the primary and secondary rainbows. They are part of the bows, but they are not features caused by refraction. The supernumerary arcs are an effect of *diffraction* and thus in the family of *coronas*, which we will explore in Chapter Two. The physics of the bow itself was basically known by the end of the seventeenth century, but a proper account of these little arcs was not available until much later.

The two bows and the supernumerary arcs vary greatly from one display to the next, so if anyone could ever tire of seeing such colors in a perfect curve in the sky, he would still have surprising differences to look forward to each time. But a large part of the attraction of rainbows comes not from their love-

---

* See the Annotated Bibliography, pp. 281–287, for this and other useful books and periodicals mentioned in the text.

liness and splendor, which are undeniable, but from their strange nature. Even though we can trace the path of light rays through their reflections and refractions and understand how the rainbow operates, the phenomenon strains the capacity of ordinary language to discuss it. A rainbow is not an object in the ordinary sense of the word. We expect an object to occupy a position in space. Like objects (the sun, moon, and stars) that are so far away that they may be considered to be effectively at infinity, a rainbow follows an observer as he moves. But the raindrops that cause a bow are not necessarily far away. Out in the Midwest United States a friend of mine saw a section of rainbow on a storm that was 80 miles distant, but Minnaert (p. 169) says that the rain causing a bow is seldom more than 1½ miles away and sometimes very close indeed; he mentions an observation of a rainbow in front of a wood just 3 yards away! That is an unusual case but, as we shall see presently, fog-bows can often be even closer. Should we say the rainbow is located where the drops forming it are? If anything, it would be best to say that the rainbow is all of the light that is emerging from the drops at a certain angle. Thus the rainbow is *light*—an addition to the brightness of the background—coming to us from a certain preferred direction.

The best way to visualize the rainbow's position in space is as a cone of light extending from the raindrops to its apex at the observer's eye. The thickness of that cone, which is hollow, is stained through from top to bottom with color that grades smoothly through countless shades from red to violet. This leads us to a further amazing conclusion: for each observer there must be a different cone producing a different rainbow. It is quite correct to speak of two people standing side by side seeing two slightly different rainbows. So a shower that gets sunlit can actually carry with it an almost infinitely rich tangle of intersecting cones trailing it in the direction of the sun. And the sights at the apexes of these cones are all different, all constantly changing with the process of evaporation, rainfall, availability of sunlight, movement of the storm, and other factors. You can see that language is not used to dealing with a thing of the rainbow's nature—the only way the rainbow can fully enter language is in a strange blaze and twist of wordage which is a counterpart of the bow's own weird glory in the sky. (By the way, the base of the rainbow's cone does not have to be perpendicular to the plane which contains the sun, eye, and rain—in other words, the rain forming one end of a bow may be much closer to the observer than that forming the middle or the other end.)

The rainbow is even slightly different for each eye of a single observer, although usually our stereoscopic vision does not perceive this. In most rainbows the droplets are far enough away for the difference in angle between the two eyes to be unnoticeable. The exception is many artificial rainbows. Artificial rainbows can frequently be seen in fountains or lawn sprinklers, as long as the observer is properly positioned and the droplets are of suitable size. In such instances, there is a sufficient number of droplets very close to the observer and the rainbow is formed close enough so that each eye perceives its own rainbow.

Almost every culture has had its lore and legends of the rainbow. The most famous tale is certainly the story of Noah, who saw the rainbow that marked the end of the biblical deluge. Also renowned is the goddess Iris of Greek myth, who was a messenger of Zeus. Her name has gone into English to describe the colored part of the eye (*iris*), and to evolve into the adjective *iridescent*, which means "rainbow colored," and may be used to describe everything from certain fishes' scales and snakes' skins to certain beautiful clouds (which are discussed below in Chapter Two). In Norse mythology the rainbow was the mystical bridge Bïfrost, which the gods traveled on to their home, Asgard, and which was watched over by the god Heimdall, who would blow his horn and destroy the bridge on the day of Ragnarok, when the forces of evil would wage their world-ending attack.

There are many tales of the rainbow from other cultures, and some of them are as rich and imaginative as the Greek and Norse. One people in South America says the rainbow is a monstrous serpent which once grew out of control until it stretched across the sky. The serpent threatened both heaven and earth, and all would have been lost if help had not arrived at the very last moment. It was the Army of the Birds that came to the rescue. Flocks of all the different kinds of birds, save a few, soared in by the billions, tearing and pecking at the monster, finally slaying him. His blood of many colors flowed over his body and his attackers, so that the birds, which until then had all been very dark-hued, became bright with all the colors. And so the beautiful plumage of the birds (except those few dark kinds which did not help) became forevermore a reminder of the great deed they accomplished on that day they saved the world.

Clearly, even the rainbow with which we thought we were familiar is more varied than we imagined: it has its higher-order relatives, its supernumerary arcs, its endless batch of cones. But there are many more different versions of the rainbow phenomenon, which vary with the light source and the medium that does the reflecting and refracting. Not all of these other, little-known bows are uncommon. To observe some of them, one only has to know where and when to look.

One kind of rainbow which will never be a common sight, however, is the *moon rainbow*—a rainbow with the moon, not the sun, as the light source. How beautiful it must be to see a rainbow in the dark with perhaps a few stars peeping through the clouds on which the bow appears. Unfortunately, the moon rainbow lacks much of the splendor of its daytime counterpart because it is so much dimmer and therefore its colors are weaker or undetectable (the color receptors in our retinas do not register the impression of color unless the light is sufficiently bright). The moon is immensely fainter than the sun—so much fainter that only near the full phase is a lunar bow possible. Therefore, on only a few nights a month can a moon rainbow be seen—if all the other necessary conditions just happen to be met on one of those few nights. But the situation is not as hopeless as it sounds. On several occasions rainbows have even been observed during the total part of a solar eclipse, when they were caused by the sun's glowing atmosphere, which is seldom much brighter than the full moon. These periods when the sun is totally eclipsed usually last for only 2 to 3 minutes and occur in a very limited area of the world no more than a few times a year (usually less). If eclipse-of-the-sun

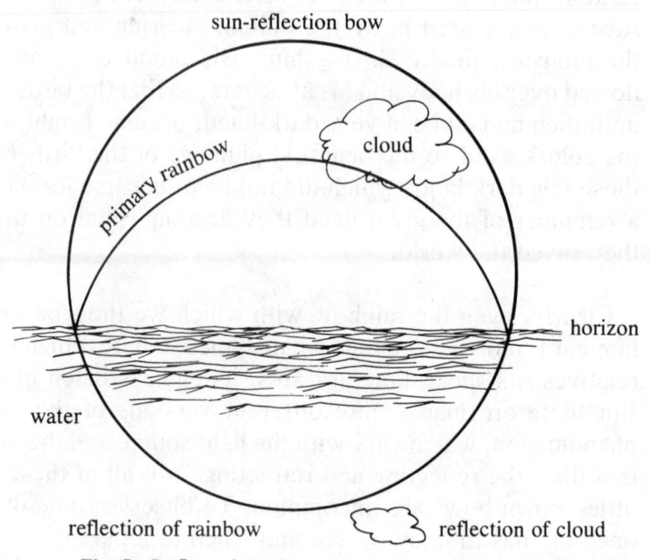

Fig. 5. *Reflected rainbows. Notice that the cloud's reflection is displaced away from the horizon but the rainbow's reflection is not (the rainbow, like the sun or moon, behaves as if it were at infinity).*

rainbows have been seen, then surely moon rainbows cannot be as rare as we might fear. The key to seeing them is being aware of the weather and the world you live in (weather is, after all, the largest part of the subtle and ever-changing mood of the natural world). Be sure not to confuse a lunar halo around the moon (a fairly common phenomenon) with a lunar rainbow, which is seen in the opposite direction from the moon.

Another light source which can cause a rainbow is the image of the sun in water. Such a rainbow is thus one in heaven caused by a light source on earth (or on water, if you prefer). Of course, the ultimate light source is still the sun, but here the trajectory of the rays is a little harder to visualize. Figure 5 makes the situation clearer and shows in a particular case where this rainbow would appear in relation to the ordinary primary bow. The diagram shows also the simpler and very common case of the ordinary rainbow's image seen reflected in water. Imagine what it would be like to see all three at once, as they are portrayed in the diagram. And suppose the secondary bow was visible with each of the three, too! Minnaert points out that one may sometimes see just parts of the rainbow caused by the sun's image in water and that this could be a result of the positioning of various bodies of water in one's vicinity. He suggests that the observer search his neighborhood to try to find the ponds that could have caused the rainbow. How beautiful and remarkable to think that ponds and puddles scattered over a landscape like a handful of jewels (each containing its own sun) could paint their own pieces of a strange rainbow in the sky, or to think that a rainbow in the sky could give us detailed information on local geography by portraying the landscape in its form!

The image of the rainbow reflected in water should, of course, occur in any sufficiently calm body of water whenever a bow is clearly visible in the sky, but though it is much more common than the reflected bow just described, this phenomenon is, in several ways, even more thought provoking and strange. Of course, as with all reflections, the colors in this reflection of the rainbow seem warmer, richer. But notice also that the rainbow's reflection in water appears lower in relation to a cloud's reflection than the rainbow does seen against the same cloud in the sky. Since the bow behaves like an object at infinity, its reflection does not appear shifted farther from the horizon, as do the reflections of objects like clouds, hills, or trees. A little further thought, however, brings us to a truly amazing realization: the image we see in the water, and assume to be the reflection of the rainbow we are viewing in the sky, is really *not* that reflection. It is the reflection of another rainbow, a rainbow we cannot see!

To make this situation sound more puzzling still we can give an abbreviated version of the truth and say that the rainbow has no reflection! But if we want to speak more precisely and fully we should say this: it is impossible to see in a body of water the reflection of the same rainbow we are seeing in the sky. The explanation for this remarkable fact is simple if we recall that even observers standing side by side see slightly different rainbows. The image on water is a reflection of the rainbow visible from that position where the image is. And that rainbow is caused by different droplets (ones in a lower cone) than those causing the rainbow visible from shore.

This fact may seem at first to be of only technical interest—for, after all, do not side-by-side observers see rainbows so similar as to be indistinguishable from each other? Yes, but an image on water may be quite far from an observer, and there can be striking differences between the bow in the sky and what seems to be (but is not!) its reflection in the water. Perhaps the most amazing thought of all is this: it should actually be possible on rare occasions to see a rainbow in the sky fade out, leaving what seems to be its reflection still visible on the water below! That is a trick which even the Cheshire cat could not equal, but it is just one of many marvelous consequences of the nature of rainbows. (By the way, a person presumably *can* see the reflection of the same rainbow he observes in the sky by turning from the rainbow to look in a mirror behind his head.)

Rainbows are formed by droplets of at least 0.05 mm in diameter, but bows that are fairly common and strikingly different are formed in the smaller droplets of fog. Whenever fairly thick fog drifts into your neighborhood, walk into its enveloping cloak—if it is night, with your back to a streetlight or other rather bright light source; if in morning, with the low sun behind you. Suddenly you make out a white arch looming in front of you, inviting or daring you to enter. It seems just a few feet ahead, and indeed it can be seen in front of objects which are that close, so the tiny fog droplets causing the ghostly apparition must be no farther. But as you step toward the arch, it recedes at precisely your pace—walk or run. If it is night you must not pursue it too far from your streetlight or it will fade. It changes its intensity as the fog drifts by, though not precisely in accord with the thickness of the fog. Now it has gotten more prominent. Suddenly with a thrill you see the arch extend between you and the ground not far in front of you. You are seeing a full circle of white light, now tinged dimly with orange on the outside, blue on the inside: *Ulloa's Ring*.

Fog-bows have been remarked upon since at least the Middle Ages, but the most famous early observation was made by Captain Antonio de Ulloa and his party (including Pierre Bouguer) when they looked down from a mountain in the Andes on fog below them just after sunrise. They saw their long shadows on the fog with a circle of light centered on the shadow of just one person's head—the observer's own, of course, for from each person's own viewpoint the shadow of his own head marked the antisolar point about which the 360° fog-bow curved.

As the droplets get smaller, the angle of the fog-bow from the antisolar point (and hence the radius of the circle) begins to diminish, so the primary fog-bow is located at less—sometimes much less—than the 42° which is always the angle of the primary rainbow. On the other hand, the width of the primary

Fig. 6. *Fog-bows can sometimes be produced by sun on fog or (as in this photograph) by streetlights or car headlights on fog. In this view, the primary is represented by the outer edge of the illuminated area; supernumeraries blend into its inner edge; a faint, rare secondary is not visible here.*

fog-bow itself is up to two times greater than that of the primary rainbow. Another major difference is that the first supernumerary of the fog-bow is quite prominent when it occurs and is seen far more often than the secondary fog-bow, which, according to various authorities, is very rare. The best fog-bow I have yet seen occurred on a night after 11 straight thunder-days (days on which thunder is heard) and had as its light source a streetlight. I later enhanced it with the high beams of a car, but with the streetlight alone and the dark tree-lined country road I was able to see the full circle, sizable portions of two supernumeraries, and a large part of the secondary. The bottom of the circle appeared as a bright band running across the road in front of me. Most astonishing of all was the view when I lay down on the road, for then, from my shifted perspective, the bright band, the bottom of Ulloa's Ring, ran across the bottom half of my body: a fog-bow was lying on me; I was "touching" a fog-bow!

Displays of the fog-bow as prominent as that one are probably quite uncommon, but, then again, how many people have any idea that there is such a thing as a fog-bow (it is much harder to discover a thing by accident)? It is certainly not too uncommon to be able to see a dim fog-bow as you walk away from a streetlight (you must have a dark background to view the bow on, however). Since fog is only a type of cloud, it is also possible to observe cloud-bows from an airplane, from which an even more beautiful (though perhaps not so eerie?) phenomenon may be seen directly around the antisolar point (and thus at the center of a cloud-bow circle): the *glory*, which we shall encounter in Chapter Two (glories are also visible, though not so easily, in fog). Airplane flights are a fascinating opportunity for many unusual observations of the sky and earth, but for most of us the more frequent (and less expensive) opportunities to see cloud-bows will be in the clouds on the ground, fog. The fog need not be thick—there is even a report of a fog-bow being seen when the observer could not detect any fog! A friend of mine described riding a bike at night past each succeeding streetlight, zooming toward and chasing one suddenly looming fog-bow after another. Whether you seek fog-bows night or day, the fog is a mysterious place, a world of its own, and one of the strangest marvels you may come upon inside of it is Ulloa's Ring.

How about a bow which lies on the ground curving in toward the observer as a hyperbola with ends receding into the distance? Such is one form of the *dew-bow*, formed in dew on grass and also sometimes in other situations. It has the rainbow colors and is best seen when the sun is low. Secondary and reflected dew-bows are also observed.

There are even stranger bows. Can bows be formed from any substances other than water? Yes, this is proven in laboratory experiments. Even more interesting: is there a rainbow at wavelengths other than those of visible light? Yes! A part of the rainbow should occur in the near infrared range, those wavelengths just a little longer than the ones of red light. To prove the existence of the *infrared rainbow*, Robert Greenler of the University of Wisconsin used film sensitive to those wavelengths and succeeded in obtaining a photograph of that bow which no man had ever seen. But some animals can see into the infrared spectrum. Could snakes possibly have any awareness of rainbows? It has been proven that some species of birds migrate with the aid of very specific groups of stars. A bird called the Indigo Bunting may have been aware of the stars that form the Big Dipper long before any men lived in the New World, perhaps even before there was a species of man or the star pattern took its present form.

Last of all, let us return from these strangest bows (and another connection between birds, snakes, and rainbows) to consider the primary rainbow in special settings that make it all the more uncanny and lovely. Not to be confused with the infrared rainbow is the *red rainbow*, which is the answer to the question, what does a rainbow become when the sun sets? The other colors are more strongly scattered and absorbed; thus just as red is the last color left to predominate in the setting sun, so too red predominates in the rainbow. As sunset approaches, the primary rainbow is now well up in the east, huge, and its colors are fading out one by one. But even for minutes after sunset the red rainbow may remain visible because at the higher altitudes of the raindrops the sun is still shining even when the observer on ground level is already in twilight. The ends of the red rainbow gradually dwindle until only a section and then a single spot of red is left, glowing brightly over the observer in the dark . . . and then is gone.

There are many stunning settings for ordinary rainbows. One may see the very top of the rainbow's arch rise in the east as the sun reaches an altitude of 42° from the western horizon. Rainbows usually occur on the back edge of storms—partly because of the typical structure of storms and partly because the abrupt showers which are likely to produce rainbows are more likely to occur in the latter part of the day. It would nevertheless be possible to see a storm advancing with a rainbow leading the way (I know someone who has observed this). But surely such an observation is very rare: consider the rainbow's reputation for peace, which exists to a certain extent because a rainbow almost always occurs at the end of a storm, not before. Practically as uncanny as a rainbow leading a storm is a rainbow hung over a snowy landscape; yet that strange sight has been both seen and photographed (Robert Greenler managed such a photograph with the help of spray from Niagara Falls in winter, but rain after a snowfall would suffice). In the Midwest United States I finally succeeded in seeing lightning striking apparently across (but maybe really *through*) a rainbow.

Of all the endless possibilities of rainbows seen against unusual and beautiful backgrounds, the favorite I have seen is a rainbow in front of distant trees which are at the height of their autumn blaze of color. One autumn in New York State I observed a rainbow whose one end passed in front of a high hill about a mile away. The hill was covered completely with orange and red trees. I could see the individual trees through the rainbow quite clearly. It is always surprising how transparent the rainbow is, but we should not be surprised, considering that it is only light. The trees were breathtaking viewed through a mile's thickness of red light, but even more astonishingly beautiful were those red and orange leaves shining through the curtain of the bow's blue and green. Such wonders draw us out of our common selves to what seems another world but is really a larger world of freedom that we can help ourselves grow to. It is not beyond but on the road of the rainbow that we come to that larger realm.

CHAPTER TWO

Halos, Coronas, and Glories

Although there are many more kinds of rainbows, in more places than almost anyone would imagine, a hard fact remains: in the course of a year, even an observant and fortunate person is unlikely to see more than a few displays of the primary rainbow in the sky. That seems like a severe limitation on color and loveliness in the sky, and indeed it would be if there were no other phenomena which could produce strong-hued and marvelously regular patterns in the heavens. But there are! There are especially those phenomena, or families of phenomena, which are among the most beautiful and prominent in all nature and yet, strange to say, among the least popularly known. I am referring to those grand and extensive displays of patterned color in the sky called *halos* and *coronas*.

Of all the dozens of halo phenomena, perhaps only one, the 22° lunar halo, the "ring around the moon," is known at all to most people. There are even many astronomers and meteorologists who know little about halos or coronas, and who are surprised to learn that there are whole families of phenomena which can sometimes rival even the rainbow in beauty, yet include members which are clearly visible dozens of times a year. How is it possible that halos and coronas are so lovely and striking and yet virtually unknown to the majority of people on Earth? One good reason is the fact that many of these displays are brightest when found in the vicinity of the sun. The commonest and many of the most prominent halo phenomena are easily far enough from the sun not to dazzle the eyes, but people usually avoid looking even in the general direction of the sun unless they have a special reason to do so. Halos, coronas, and iridescent clouds (which are detached portions of a corona) are more than sufficient reason to look in the general vicinity of the sun. In this chapter we will take a brief survey of these rich and varied sky effects, and of the related phenomenon called, simply and appropriately, the *glory*.

All of us have heard that "a ring around the moon means rain soon." The role of halo phenomena as precursors of bad weather is one that has been long and widely maintained. The Zuni Indians have a saying that when the sun is in his house (a halo around the sun), it is going to rain. Weather lore such as this is quite often proven correct because halo phenomena occur in cirrus and cirrostratus clouds, types frequently seen as the forerunners of the vast storm system and rain area which can accompany a warm front.

Halo phenomena are caused by the refraction and reflection of light through the faces of the countless little hexagonal ice crystals that make up these clouds. The most important types for halo formation are long *pencil crystals* and squat *plate crystals*. Sometimes the crystals assume preferred orientations as a result of aerodynamic forces; these regular positionings serve to create more of the dozens of different halo effects—colorful and graceful circles and arcs, pillars, and spots, which can appear in many places in the sky.

The most famous of the halo family is certainly the *22° halo*, so named for its radius of about 22° from the sun or moon at the center of its circle. It is sometimes also called the *small halo*. This may surprise you if you recall from Chapter One that your fist held at arm's length subtends an angle of about 10° and the moon or sun about 0.5°: the "small" halo has a diameter that can reach from the horizon to halfway up the sky. The *large halo*, which is almost never seen in its entirety, has over twice the diameter, yet is still not the most extensive of the halo phenomena that you may observe.

The 22° lunar halo is the "ring around the moon," but the stupendously greater intensity of sunlight creates a halo which is far brighter and more colorful. Whenever cirri feather a large area of the sky around the sun, scan the region at a distance of 22° from the sun for traces of the circle. The halo forms at 22° because that angle is the minimum angle of deviation for

Fig. 7. *A complete 22° halo. The bright area at the center is the several-degree-wide region around the much smaller disk of the sun (this area overexposes the film, making the disk itself undetectable in the photograph). Photographed in Wisconsin by Robert Greenler.*

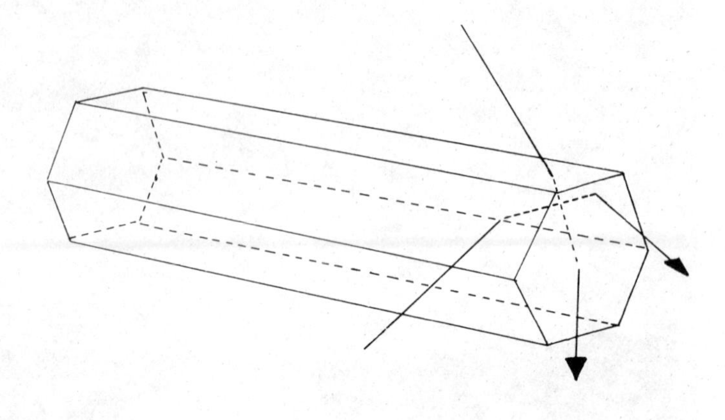

Fig. 8. *Paths of light rays through a long hexagonal crystal.*

light which passes through faces on alternate sides of the hexagonal crystals. As in the case of the secondary rainbow, the sky is darker inside the halo than outside, and the reason is the same: some light is deviated by more than the critical angle, but none at less. As a result, the halo has a sharp inner edge and a much more diffuse outer border. The crystals which cause the 22° halo have no preferred orientation; thus all orientations are represented and the halo appears as a concentration of light at 22° from the sun in all directions—a perfect circle of radiance.

The 22° halo is remarkably common, though we are far more likely to see just sections of it than the entire grand ring. At least parts of this halo can be spotted on an average of perhaps one out of every five days in most climates at any time of year. Some observers have seen it on more than 200 days in a single year! The band of the halo, which is about 1.5° wide (comparable to the width of the primary rainbow), appears colorless on many occasions, but often one can detect red on its inner edge and sometimes blue on the outer. Red is the most common color in this halo because its wavelength is longer than those of other colors and so it is the only color which is not overlapped, appearing on the inner edge (light is deviated at not just the minimum angle, which is slightly different for each color, but at greater angles in diminishing intensity). Even red is not often seen in lunar halos because of the weakness of moonlight.

The other kinds of halos proper (that is, circles of various sizes with the sun or moon at their center) are all rather rare, so first let us consider some very common halo phenomena which are quite different. The most common and one of the most striking of these are the *mock suns* or *parhelia*.

Mock suns are concentrations of light which appear as roughly elliptical (or sometimes diamond-shaped) patches at the same altitude as the sun but always at least 22° away from it on either side. They are the result of light refracted through faces on alternate sides of the crystals, but, unlike the 22° halo, through crystals which are all oriented with a vertical central axis, so that with increasing elevations of the sun the parhelia appear at ever greater angles to either side of the sun. It is only when the true sun is on the horizon that the mock suns are exactly 22° from it; at higher altitudes of the sun, the parhelia move farther out; when the sun is 61° or more above the horizon, the mock suns are no longer visible. The 22° halo is not always seen with the parhelia, nor need both of the parhelia be visible; a person may sometimes even see a small solitary wisp of cirrus that is passing by one of the parhelion positions be suddenly kindled into a blazing patch of colors as the tiny ice-prisms focus sunlight to the observer.

Just how bright and vividly colored the mock suns can become is astonishing. A parhelion may sometimes be too radiant to look at, and I know of one observer who has seen his shadow cast by a mock sun! When their brightness is less fierce, the colors are sometimes unbelievably saturated. I have seen reds so deep and intense that the sky looked as though it had been painted—but with glowing pigment. Unlike the 22° halo, the mock suns may show all the spectral colors, and quite purely. The colors most often seen are red (on the inside, facing the sun) and blue (on the outside, away from the sun). Minnaert (p. 196) implies that yellow is also seen reasonably often between these two colors. I agree with him in that yellow seems to me the third most common hue in parhelia. The yellow in mock suns is especially delightful because almost nowhere else in sky phenomena do we get to see this color purely (sunlight is yellowish white at best).

Parhelion is Greek for "alongside the sun"; another name for the phenomenon it identifies is *sun dog*—because the parhelia seem to dog the sun across the sky. (The term *sun dog* is sometimes also incorrectly applied by people to other sky effects, such as solitary sections of rainbow, which, by the way, have themselves been referred to by the interesting term *wind galls*.) Although it probably has nothing to do with the origin of the name, sun dogs can have *tails*! The tails (this is the actual technical term) are bluish white streaks of light which extend outward from mock suns for distances sometimes greater than 20° (their length depends ultimately on solar elevation).

A very rare effect which nonetheless deserves mention here is what has often been called mock suns of the large (46°-radius) halo. It is now thought by some experts that these mysterious spots of light are a secondary halo phenomenon, mock suns of mock suns. If parhelia can be bright enough to be dazzling, then it certainly seems possible that they could cause secondary mock suns. After all, *parselenes* ("alongside the moon")—mock moons or moon dogs—are not extremely rare, though they are typically rather faint and colorless and usually require a moon that is near full.

In the Hebrides Islands, mock suns are called *gaas* and it is said that "a gaa before is for a snore [hard rain], a gaa behind you need not mind." This is sometimes true because if a parhelion appears only behind (to the east of) the sun, the clouds may be only fragmentary and not the beginning of the vast cirrus vanguard of a storm system. Many farmers and outdoorsmen pay attention to the sky and are familiar with the 22° halo and with sun dogs, which are practically as common as the 22° halo.

After these halo phenomena we come to others which are still quite common, yet for special reasons have never been well known to any but the student of halos, though they are sometimes truly spectacular. One of these is the *circumscribed halo*, which, as Robert Greenler (p. 33) points out, should really be called *circumscribing*, for it does just that to the 22° circle at certain elevations of the sun. The reason the circumscribed halo has never entered into folklore or popular knowledge is no doubt its strange variance of form as a function of solar altitude. Unlike the parhelia, which basically change only their separation from the sun as it climbs or descends, the circumscribed halo goes through drastic alterations. A casual observer would never suspect that the pattern seen when the sun is 10° above the horizon is caused by the same factors as the pattern when the sun is much higher. The shapes of the circumscribed halo for various altitudes of the sun are shown in Figure 9. Note that when the sun gets very high the circumscribed halo becomes indistinguishable from the 22° one; also note that when the sun is fairly low the most prominent part of the display is the *upper tangential arc*. This arc, tangential to the top of the 22° halo (whether the latter is visible or not), is the most

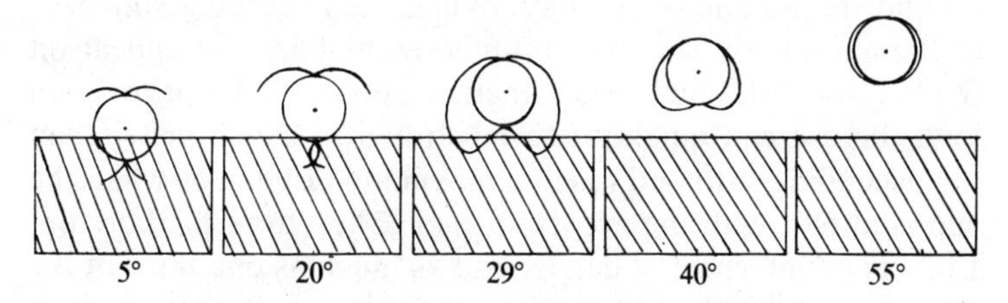

Fig. 9. *The appearance of the circumscribed halo for different elevations of the sun. Drawn by Chris Duke, after Minnaert.*

frequently observed section of the circumscribed halo and may often burn intensely with many or all of the colors of the spectrum. It is not too unusual to see even awesome oranges and greens in this arc!

Of all the grand and colorful halo phenomena, the loveliest may be the *circumzenithal arc* and its counterpart, the *circumhorizontal arc*. The circumzenithal arc is visible more often than rainbows—perhaps far more often—and some observers have mistaken it for an upside-down rainbow, high in the sky. It is probably the position of this phenomenon which has kept it from being commonly marveled at and becoming a part of folklore. Not many of us make it a habit in our lives to look up in the vicinity of the zenith, whether or not there are cirri in the sky. How strange to think that we are therefore often walking around unaware of one of nature's loveliest sights hanging right over our heads!

The circumzenithal arc only forms when the sun is 32° or less above the horizon, and for that precise altitude of the sun it appears as a beautiful flushed spot of color right at the zenith itself. As the sun descends the spot opens out like a strange and wondrous flower into an arc which can show all the spectral colors with deep saturation. At a sun altitude of about 22°, the circumzenithal arc is brightest; at best, it may extend at its height for as much as one-third of the way around the imaginary circle that has the zenith as its center.

This phenomenon's counterpart, the circumhorizontal arc, can only form if the sun is at least 58° (90° − 32°) above the horizon and therefore, in the northern United States and southern Canada, is visible only during the late spring and summer, around the middle of the day. When the ascending sun rises to the altitude of 58°, the circumhorizontal arc can appear on the horizon. Although it is sometimes obscured by haze down low, the arc can quickly rise much higher (much faster than the ascending sun) and reach a maximum brightness when the sun is at 68°, with the arc itself then 22° up from the horizon. Like its counterpart, it can extend as much as one-third of the way around the sky but at a low elevation, so its arc can be a vast, brilliantly colored band. The band is also extremely wide; thus one could argue that this is the most spectacular color phenomenon in the heavens. It is unfortunate that the circumhorizontal arc is of quite infrequent occurrence.

All of the halo phenomena so far mentioned involve refraction, but there are many other fairly common types which are caused by reflection only, and so do not show spectral colors. Before we turn to this class, however, there is still a group of rare effects involving refraction which deserves mention: the relatives of the 22° halo, the other halos proper. These are not close relatives—they do not have identical or very similar means of production—but all of them appear as circles with the sun at their center.

The *46° halo* or *large halo* has very seldom been seen in its entirety, which is not surprising considering that the sun would have to be at least 46° high and cloud conditions would have to be suitable all along its circumference—a ring which can curve from the horizon to past the zenith. On the other hand, a very enthusiastic observer (who knows where to look) might glimpse a piece of this giant ring more than once a year. The reason that even parts of the 46° halo are seen far less often than the small halo is the former's lesser intensity, for the 46° halo is formed by the smaller amounts of light which pass through one end face and one side face of the crystals. The large halo can have an intensity only about one-tenth that of the small one, though of course variations in the cloud cover can cause it to look relatively brighter at a given time in a given display. While the radius of the large halo is just over twice as great as that of the small halo, it is also true that the band of the former is, coincidentally, a little more than twice as wide. Although the diffuse outermost edge of halos will often make them seem wider, theoretical calculations give the widths of the bands of the 22° and 46° halos as 1°27′ and 3°08′, respectively. (In angular measure, each degree is divided into 60 minutes of arc [′], and each minute of arc into 60 seconds of arc [″]. Whereas the moon is about 0.5° or 30′ in diameter in our sky, the disk of a planet like Jupiter is never more than about 50″ in diameter and requires a telescope's magnification to be seen as more than a point of light.)

Along with the more common halo phenomena, which are a source of constant (or at least weekly) satisfaction to sky-watchers, there are the rare halo forms which perhaps will appear, virtually without warning, only a few times in the life of a patient observer, to reward him magnificently. When these rare types become visible they are usually parts of vast and intricate *halo complexes* in which the common effects may themselves sometimes blaze with their greatest intensity. The polar regions are most highly favored for such displays, but they are possible anywhere that cirri are seen.

Among the most exciting of the rare halo phenomena are the *halos of unusual radius*. Although they are seen far less often than parts of the 46° halo, the halos of unusual radius occur in conditions which usually give rise to more than one of them at a time. It is still not known exactly what crystal forms produce these effects, but it is believed that ones with pyramidal ends could be responsible. The question reminds us how marvelous it is that a white, feathery cirrus cloud is a swarm of tiny hexagonal ice-prisms, sometimes drifting in alignment, and sometimes focusing sunlight in the atmosphere into colored concentric circles and many other kinds of arcs and spots.

We do not know the size of all the possible halos of unusual radius, but there are several of them whose existence and approximate radius have been well established by at least a few careful observations and photographs. Halos with radii of about 8° and 17°–18°, and one with a radius between 30° and 35°, have all been reliably sighted, and on a few occasions as many as six halos proper have been seen at one time! By far the best-documented of such displays is the one which was visible over an area extending for at least a few hundred miles, from central southern England to the Netherlands on the afternoon of April 14, 1974—an Easter Sunday. This magnificent complex was at its best late in the day and for at least one observer included complete and nearly complete upper halves of halos with radii of about 8°, 18°, 19°–20°, 22°, 23°–24°, and 32° (but no trace of the 46°). Yet even this series does not exhaust the full list of halos proper which have been observed. The one which may be the most awesome of all we will discuss in another context in a moment, after we have considered the common halo effects which result from reflection only.

The most frequently observed reflection phenomenon among halo effects is probably *sun pillars*. At least a few times a year an observer who often looks at the sun when it is low will notice a beautiful feather or column of light, usually extending for a few degrees straight upward from the sun. A pillar sometimes also extends downward from the sun, though these are rather uncommon and usually shorter. The pillars vary greatly in intensity from one occurrence to another and may occasionally be astonishingly bright and striking. One of the best sun-pillar displays I have observed occurred in the clouds preceding a major northeastern United States snowstorm a few years ago. The pillar below the sun stretched for as much as 5°, while above the sun a pillar extended for a full 22°, where it was crowned by a fine upper tangential arc (a phenomenon which occurs not infrequently with sun pillars).

For a long time it was thought that sun pillars could be caused only by the squat plate crystals, floating like mirrors with their broad sides horizontal. This explanation suffices for pillars observed when the sun is very low or just below the horizon, but it cannot account for the ones which, though rather short, are sometimes most definitely observed at sun altitudes as high as 20° or more. It was not until about ten years ago that the problem was solved and much more was learned about the pillars, by Robert Greenler and his colleagues at the University of Wisconsin. These researchers investigated the formation of sun pillars by using computer simulations and discovered that the ones seen at higher solar elevations (as well as some of the low-sun pillars) must be formed by pencil crystals.

The work of Greenler and his associates also demonstrated the relationship between pillars and a remarkable halo effect called the *subsun*. The subsun can be observed from airplanes or mountains as an eerie, elliptical, sometimes dazzlingly bright patch of light on clouds below. It surely accounts for many of the sightings of UFOs from aircraft, and yet it is really nothing more or less than a reflected image of the sun on a cloud. The computer simulations show that a pillar seen far enough below the horizon shortens into an elliptical spot; the subsun is formed by ice crystals floating like a sea of almost perfectly horizontal mirrors. (It is sometimes possible for an observer on the ground to see the subsun in a valley below, or to see below his horizon a pillar which is then called a *sunstreak*.)

The wedding of computer and halo effects may seem a strange one, but it has yielded important insights into halos which otherwise would never have been possible. The research on sun pillars was just the beginning of computer simulations of halo phenomena; in the years since then, Greenler has conducted many more such studies of the various halo effects, often with exciting results. And for our interest here, one of the most notable things about Greenler is his motivation: like Minnaert, he seems to have undertaken the study of meteorological optics as a labor of enthusiasm and fascination, often

Fig. 10. *Numerous light pillars extend far up from the lights of a nearby town in North Dakota, in this photograph by Steve Albers taken on February 25, 1979—the night before a total eclipse of the sun was visible from parts of the state.*

referring to his original inspirations by these sights and his wonder in beholding them. Both Minnaert and Greenler are strong evidence that intelligence and work are not the only essential ingredients of a great scientist; enthusiasm and a sense of wonder are required as well. Those last two qualities are certainly crucial ones in making the full person of which the scientist is only a part.

The sun is by no means the only source which can produce pillars. *Moon pillars* are observed fairly often, and as far south as the northern United States it is occasionally possible to see *light pillars* associated with artificial lights, especially streetlights and car headlights. These light pillars, however, are only observed when crystals themselves are near the surface of the Earth. This is an unusual form of snow (if indeed it deserves that name), for what we normally see in a snowfall are more complex flakes or aggregates of flakes. The simple crystals are very small and supposedly only common in the very cold climates of far northern lands and high mountaintops; snow aggregates, as we have all seen, can sometimes be very large indeed: the record for any storm is one which was composed of aggregate snowflakes measuring 15 × 8 inches! While we are on the subject of snow, it is interesting to note that the 22° and 46° halos can sometimes be observed on fallen snow as hyperbolas in the direction of the low sun, and Minnaert (p. 207) reports that various halo phenomena have been observed in whirling snow—but surely this observation must have been made in a fall or mist of simple crystals?

Whereas pillars of the sun and moon behave as if at infinity, light pillars can change their appearance greatly according to how close the observer is to the light source. If one is close enough the effects of perspective can cause light pillars to appear to converge at the zenith! Greenler (p. 73) mentions a case in which a display of light pillars was probably what people in a northern town thought was an excellent aurora. In the next chapter we will see how this mistake could have been made.

Another reflection phenomenon which an interested observer can sometimes spot is the *parhelic circle* (also sometimes called the *horizontal circle*). The full circle is certainly not visible very often, for this is a ring which goes around the entire sky at the height of the sun. Not only the sun and the parhelia appear to lie on the parhelic circle. In superb displays, the *anthelion* and the two *paranthelia* may be observed as white spots like large beads on the circle. The anthelion ("against the sun" or "opposite the sun") is not located at the antisolar point, which is below the horizon when the sun is in the sky; it is found at the same altitude as the sun on the opposite side of the sky—in other words, 180° of azimuth (horizontal measure) away. The paranthelia are concentrations of light located 60° of azimuth to either side of the anthelic point. When searching for a sign of these spots, remember that 60° of azimuth does not necessarily correspond to angular measure such as we have been doing with our fists at arm's length. If the sun is low in the sky, the parhelic circle would measure a lot longer than it does when the sun is high, yet in both cases would extend around 360° of azimuth (a full circle).

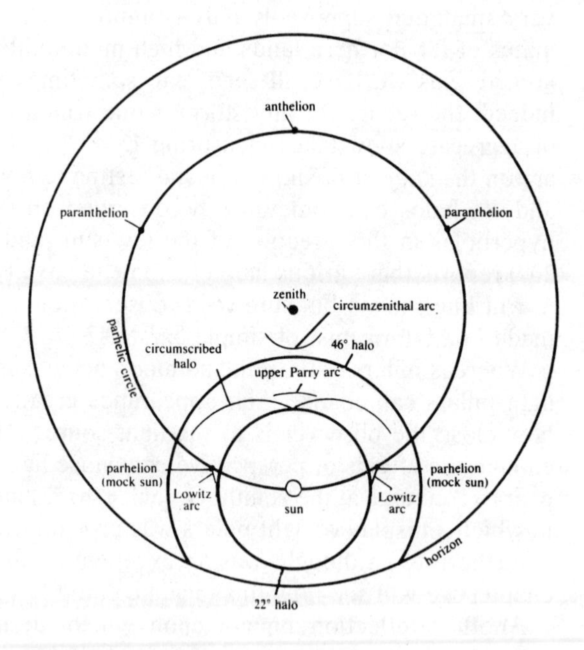

Fig. 11. *Some members of the family of halo phenomena.*

Many times the section of the parhelic circle nearest to the sun is not visible because the sun is too bright; always, however, it should appear as though the completed line would pass through the sun's disk. Or almost always: Minnaert (p. 205) mentions a strange case in which the circle ran 1° to 2° below the sun, and I bring this possibility to your attention because of an unusual observation of my own. It was a cool August day by the ocean, perfectly clear until the first cirri began to drift in during the late morning. The 22° halo formed and was complete, brilliant and colorful at times, but it was around noon that one of my friends suddenly spied the start of another ring. In a matter of seconds we saw the ring with the zenith at its center extend all the way around the sky except for a small break in the vicinity of the sun. But the completing section would not have passed through the sun, as the parhelic circle should; it appeared (by our independent estimates) that it would pass several full degrees above the sun. Roughly the upper three-quarters of the 22° halo was simultaneously visible: the sun was enhaloed and linked halos held the zenith. The somewhat larger, mysterious ring had been quite prominent, but after no more than two or three minutes it was quickly fading, and then vanished back into the very thinly cirrus-filled blue sky, perhaps never to be seen again.

Was this really the parhelic circle? If so, how could it have appeared displaced by several degrees? Making estimates of halo phenomena can be tricky, but in this case there seems very little chance of any error. Perhaps this was an extremely rare halo phenomenon, never knowingly observed before? In this instance, the possibility must remain doubtful, but there certainly are possibilities for marvelous discovery by amateur observers with little or no more knowledge of halo phenomena than the information provided in this chapter. And although a descriptive name for a halo phenomenon is certainly the most practical nomenclature, there is a tradition of associating a first notable observation of a halo phenomenon with the person who made the sighting (thus we hear of the Parry arc and sometimes even the circumzenithal arc of Bravais). What an honor it must be to have one's name forever attached to one of these rare, elusive, and delicate arcs, spots, or rings.

There are many less-exotic but still valuable contributions which an amateur observer or photographer can make to the study of halos. For instance, a spectacular halo phenomenon which is still not definitely explained but which, I suspect, occurs surprisingly often is the *double sun*. According to Minnaert (p. 203), it is a reflected image of the sun's disk seen just one or two degrees above the real sun and very rarely below it. Minnaert suggests that it is no more than a part of a bright sun pillar enhanced by "unequal distribution of the clouds." My one truly striking observation of the phenomenon, and the sightings of it in less impressive form by friends, all suggest to me that this is the proper explanation, but the image's incredible likeness to the true sun makes me wonder if there is not more to it. In the excellent instance I observed there was a long bar of cloud about 1°–1.5° thick separating the only very slightly irregular image from the true sun below. Just like a sun pillar, the sun image had no color of its own but took on the color of the sunlight causing it, which was in this case deep orange. That second "sun" was only a little less bright than the true one and almost as impressive as the countersun I discuss in the next chapter. But how could a section of a basically column-shaped pillar appear circular? And since there was no bar of water cloud right above the image, why was it so sharply limited on top? The double sun is clearly a phenomenon which needs further observation.

Another reason it is important that many amateur observers watch for halos is the fact that it would greatly increase the number of sightings of halo complexes. Those wondrous complexes are rare, but they are almost surely not as rare as the records seem to indicate, because so many have passed by without being comprehended. By reading this chapter and beginning to observe the common halo effects, you enter a fellowship of knowledgeable halo observers which has amazingly few members in the entire world. You can quickly gain the observational experience which will enable you to record properly the full details of a halo complex, if one occurs in your skies, using this and, I hope, other books as an aid.

What kind of sight is a great halo display? Such a complex was the mighty one that appeared in the skies of Gdansk on February 20, 1661, and was carefully observed by the famous astronomer Hevelius. Among the many features intricately patterning the entire sky that day were two giant arcs believed to have been sections of the largest halo proper of them all, Hevel's Halo, a ring with a radius of 90° (a radius as great as the distance from horizon to zenith)! Such a halo is literally sky-embracing, containing an area as large as an entire hemisphere of the heavens. This halo display also presented "the Seven Suns of Hevelius." In addition to the true sun, the parhelia, the anthelion, and a bright patch on the upper tangential arc, there appeared not ordinary paranthelia but other strange spots of light on the parhelic circle at the points where the arcs of the 90° halo intersected that circle. These spots were located at 90° azimuth from the anthelion and sun, right and left. And so at the top of the sketch he made of this awesome display, Hevelius wrote the words *Septem Soles*, "Seven Suns"!

Fig. 12. *A fine halo display, including the 22° halo, two mock suns, the parhelic circle (narrow band running through sun and mock suns), the upper tangential arc (at top of halo), and the circumzenithal arc (highest in photo). Photographed in Antarctica by Robert Greenler.*

Coronas are the colored areas of light, usually no more than a few degrees in radius, which are arranged concentrically in alternating bands of blue or green and red about the sun or moon. They are perhaps even more neglected and less popularly understood than halos, with which they are often confused. Even rather experienced astronomical observers sometimes mistakenly refer to a prominent corona as a halo, yet the differences in both appearance and cause between the two are profound.

Anyone who has watched the entrancing and mysterious passage of clouds around and in front of the moon has seen various displays of delicate colors about the bright face. But how many people, even experienced observers, can remember just what those hues were and how they were distributed, or know what caused them? Most of the time the corona-producing clouds are scattered and may be fairly swift, so the colors come and go and vary somewhat from one cloud to the next and the colors' arrangement is not always obvious. In favorable periods, however, one can clearly see that blue or green appears nearest the moon and is bounded at a distance usually no more than a few moon diameters by a ring of more prominent red. This entire arrangement is sometimes called the *aureole* of the corona, the first set of colors, in many cases appearing only as a pale, whitish glow about the moon. But in the better displays, one can observe further and vivid sets of blue or green and red—as many as four total sets ("the fourfold corona") have been observed, sometimes out to 10° or even farther.

What is the explanation for these lovely, often pastel hues with which the moon is adorning the clouds about it? Coronas are not caused by refraction in ice crystals, like the colors in halo phenomena, but rather by *diffraction* from water droplets or, occasionally, ice needles. These droplets and needles are so small that they are comparable to the size of light waves and therefore deflect the light in such a way as to reinforce the light in places while eliminating it in others. The size of the waves differs for different colors and therefore the reinforced rings of light for red and blue are found in different positions in the corona. The smaller the size of the diffracting droplets, the larger the orders or sets of colors will be. The more uniform the size of the droplets, of course, the purer the colors are, and also the more likely you are to observe a greater number of sets. If the droplets range slightly in size (not perfectly uniform) they will produce patterns of slightly overlapping colors, forming pastels and other hues such as pale green and orange.

Minnaert (p. 218) says that the coronas produced by ice needles are of lovely colors, purer than those produced by droplets, presumably because the needles are of more nearly uniform thickness. It is often difficult to tell whether water droplets or ice needles are the cause of a particular corona, but, according to Minnaert, the former have sets beyond the aureole that are only 20 percent as wide as the aureole, whereas the latter have all sets (including the aureole) equal in width. The most unusual corona I have ever seen was apparently caused by ice needles at a total eclipse of the sun and is described in Chapter Five. (It is worth noting here that one must not, in discussion or reading, confuse the corona caused by clouds with the solar corona, the outer atmosphere of the sun.)

Coronas around the sun—cloud-coronas, that is—are visible far more often than ones around the moon, but the sun is so bright that they are seldom *directly* visible. The natural and very simple way to make these intense coronas available is to look for them around the reflection of the sun in calm water. The difference is remarkable: a corona undetectable or barely perceivable through direct (and dangerously dazzled) vision becomes a prominent blue and red (and often wavelet-rippled) rose of beauty about the sun's image in the water. Many a poet and observer of nature has marveled at the entire world which seems to lie in the thin plane of a pond or a pool's surface. But of all things about that gently shaken world of rich, warm hues, perhaps the most wondrous is that one of the heavens' loveliest and most colorful phenomena is seldom seen naturally by any other means than this surface glass of ponds and humble little puddles.

We can contrive other ways, of course, to get good views of coronas. An excellent method is to look at the sun's reflection on black glass. On rare occasions clouds are just thick enough to permit a good view of a corona by direct vision, but I cannot stress too strongly the great danger of looking so closely in the direction of the sun. *It is better not to take even the slightest risks!*

Coronas are visible in reflecting surfaces like water on virtually every day that is not overcast, for they can be caused by almost any water cloud which is not too thick to obscure the sun completely. But coronas do not only occur in clouds. Another place where you may frequently observe them is on a frosty windowpane through which the moon or a streetlight is shining. *Entoptic coronas* are various coronas produced not on a frosty window of glass or the frosty "window" of an ice-cloud-clad sky, but within the eye itself! Tiny mucous particles on the surface of our corneas and radial fibers in the lenses of the eyes are probably two of the important causes of entoptic coronas. The easiest to observe is perhaps one that appears around a streetlight at a fairly short distance. How do we know that such a corona is not being caused by a very thin mist of water droplets or ice needles in the air between us and the light? We can always distinguish between a corona formed inside the eye and one formed outside by obstructing the light source. If you cover the moon with your finger, you can still see the cloud-corona surrounding it, but if you block any light source (moon, streetlight, or other) from your view an entoptic corona will disappear.

When detached areas of a corona are seen as bands, spots, and strips of the corona colors on clouds, the phenomenon is known as *iridescence*. *Iridescent clouds* may be observed by sunglasses or reflection when they are quite near the solar disk, but they can be frequently spotted with direct and unaided vision when they are at considerable distances from the sun. They are most often sighted about 5° to 20° (seldom more than 30°) away from the sun. For those that are rather near the sun, one notices that a certain angular distance is critical to the iridescence: a cloud will look quite ordinary until reaching the proper spot (where a coronal ring would be bright if the cloud cover were complete) and then will transform into blue and pink as suddenly and decidedly as litmus paper dipped in a base and an acid. At greater angular distances the size of droplets in various parts of a cloud is more important. Droplets are, for instance, smaller on the edges of a cloud, where they are evaporating, so the colors are distributed not in arcs of a vast coronal circle but in horizontal bands or spots which seem to follow the contours of the clouds. Minnaert sometimes refers to such spots as eyes, which makes me think of a peacock's tail: iridescent clouds may not have as many colors as a peacock's tail but their vividness and delicacy are comparable, and suggest that they are some of the plumage and finery of the sky itself.

Few people would believe that even in the middle of day clouds could display such intense colors, until they see a fine iridescent cloud for themselves. Through sunglasses or through the darkened windows of a bus, the clouds may be seen on many a day. They are also very beautiful swimming on ponds and puddles in the section of reflected sky around the reflected sun, but the greatest thrill of all is being able to see them prominently with ordinary vision in the sky. How wondrous it is to behold a cloud drenched in pink and green or blue in the midday heavens! One of the best clouds for iridescence is *altocumulus lenticularis*, lens-shaped wave clouds which are common in the lee of mountains but also occur elsewhere and are very stable, sometimes actually regenerating in a stationary position (they may also form in layers to give the appearance of a pile of plates).

The most beautiful iridescent cloud I have ever seen was a strange cirrus at an unusually great angular distance from the sun—centered roughly 35° away. It was virtually the only cloud in a beautiful deep blue sky when suddenly it became soaked with the purest pink and blue (which soon after became green). I almost could have shouted out loud in my astonishment and delight at this giant, delicate, stranded brush stroke of colors, but my breath was taken away. It floated ever so lazily, alone, and miraculous in that blue sky, but it was not too long before it had passed its optimal position, was drained of all its color, and became again an "ordinary" cloud—but such is the wonder secretly potent in what may seem the most unassuming things. Never have I beheld such a beautiful iridescent cloud, and seldom indeed have I seen any of them farther from the sun, though once I actually spotted what seemed definite iridescence a full 60° away.

Although the particular cloud I just described was strange and beautiful, the most unusual and beautiful class of iridescent cloud is the special kind known as *nacreous* or *mother-of-pearl clouds*. The very definition of a mother-of-pearl cloud is based on its having a high degree of iridescence (or *irisation*, another term that comes from the name of the Greek goddess of the rainbow, Iris). True nacreous iridescence occurs only in clouds as cold as about −90° C, and becomes visible only when the sun is low or, most beautifully of all, when the hour is late into twilight. The fact that nacreous clouds are illuminated so long after sunset proves that they must be very high, at an altitude where the sun is still shining long after it has disappeared for a viewer on the ground. Nacreous clouds are usually found between 12 and 20 miles above the surface, far higher than any except the wonderful noctilucent clouds which we will encounter in Chapter Three.

The very low temperatures required to cause nacreous clouds are usually found, even at this high altitude, only in extreme northern or southern latitudes. Minnaert (p. 230) says that they are usually visible from Oslo in winter when skies are very clear, and they are occasionally sighted from the British Isles. It is not impossible, however, for nacreous clouds to occur at lower latitudes; in rare cases they have been seen as near to the equator as Ghana (5° to 10°N). So those of us who live in low or middle latitudes should take heart that we may get a chance to observe mother-of-pearl clouds from home someday, and should pay special attention to dawn and sunset skies in winter. Minnaert (p. 229) says that "sometimes the entire cloud bank is almost one colour, with spectral colours along the edges or in oblong horizontal rows, between which one can see the sky, a strange opal-coloured background." It is also said that they are sometimes bright enough to color snow on the ground.

The last phenomenon we will examine in this chapter is a beauty which was seldom observed in earlier days but which now has become available to everyone who occasionally takes an airplane flight, if he knows where to look. Even people who have never heard of this phenomenon sometimes do spot it outside their airplane window and call it a 360° rainbow, which is a possible but not very likely sight (a cloud-bow is more common). What these people have usually observed is several concentric circles of prominent red and blue or green looking somewhat like a corona but free of the sun's glare because appearing at the antisolar point. This is the phenomenon called the *glory*.

The lovely colored rings of the glory are certainly an interference pattern which can be attributed to diffraction by water droplets in clouds. Is the glory then an anticorona (the corona phenomenon formed when the droplets are reflecting back the sunlight, as car headlights or a flashlight beam are shined back by cat's eyes or the reflectors on road signs)? This model (and all others that have been proposed) has met with difficulties. Although there is a mathematical treatment of the glory which can account for its appearance under different conditions, Greenler recently stated that there is as yet no really suitable physical model to help us understand the phenomenon.

The glory remains difficult to understand, but it is not impossible to observe even if you never take airplane flights. One way to spot the glory is by looking down on a cloud from high on a mountain. It is sometimes possible there to see, when the sun is low, one's shadow as an immense and distorted—in fact, conical—shape on the cloud below. This shadow is sometimes called the *spectre of the Brocken* and the glory about the shadow's head the *Brocken bow*, because this eerie apparition was traditionally visible on the Brocken, a high peak in Germany. One of the aspects of this glory which most mystified people was the fact that when several people were present, the rings of color would be visible only around the head of the observer's own shadow. We now know, of course, that this appearance is not a sign of special powers or divine favor toward the observer but a consequence of the fact that the glory appears around the antisolar point, which for each person is at the head of his shadow.

You do not have to visit the mountains to see a glory, either. The glory is sometimes visible around the head of your shadow on fog at ground level when a light source—the sun or car headlights or a streetlight—is low behind you. Whenever you see a prominent fog-bow, look also for the glory. In the classic observation of the fog-bow by Ulloa and his companions, the glory was also very prominent at the center of the 360° fog-bow (Ulloa's Ring).

If you do occasionally take an airplane trip, try to get a seat where you will be able to observe the antisolar point. During some part of any reasonably long flight you may very well spot a beautiful glory for a while on the clouds below. Are you near enough to the cloud to see the shadow of the plane? Notice that the glory is not merely centered on the plane's shadow but on the shadow of the precise part of the plane where you, the observer, are sitting. Notice whether the glory you are seeing is bright or rather dark in its center (the existence of both types has been a feature which has caused difficulties for theorists). As always with such phenomena, try to make an angular measure—in this case, of the radius of the red rings, which are easiest to see in the glory, as they are in the corona.

There is a need for careful observations and photographs of almost all the phenomena in this chapter. Photography of most of them is not extremely difficult except for the problem that halo phenomena often occur over such large areas of the sky that ideally wide-angle or even 180° "fish-eye" lenses must be used. Such lenses are costly (though there are now inexpensive "pseudo-fish-eye" lenses), but Greenler suggests an alternative which worked well for him when he was starting years ago: photograph the image of the halo in a reflecting sphere. You may use a garden globe or, as Greenler did, the least expensive device imaginable: a Christmas tree ornament!

Anyone with a little knowledge could make a major discovery about halos, coronas, or glories. Remember especially that we need a widespread network of people who know enough about halos to notice all aspects of an unusual halo complex as soon it develops, and to observe it competently. By reading this chapter and beginning to look for the 22° halo, mock suns, and other common halo phenomena, you are becoming part of that informal network or fellowship which is needed.

The observation of these families of phenomena is also more than just an attempt to further scientific knowledge, of course. It is an enjoyment and satisfaction to have the ability to bring so much more beautiful and grand color and form into your life. People often long for new realms of experience which have been surveyed by few others before. I can think of no such realm as astonishing, grand, and prominent as the realm of halos, coronas, iridescent clouds, and glories.

CHAPTER THREE
From the Blue Sky to the Zodiacal Light

This chapter is a journey from brightest day to darkest night. We begin with the blue sky, which occurs mostly in the lowest levels of the atmosphere, where we all live; for, to alter slightly what Eric Sloane has said, we are all dwellers in the sky, if only its bottom part. We end this chapter out in interplanetary space late at night. This is not our longest journey in miles in the book, but it is our greatest transition, through numerous light phenomena of every conceivable color, shape, and motion. In the scenery of this trip we meet virtually all the major phenomena of that entire world of transition called twilight. Only meteors will be left out of our itinerary, to be dealt with in detail in the next chapter. It is generally true that we will be traveling from the more familiar to the more mysterious, but there are secrets waiting for us in every unexpected place—a strange beam of color hidden until the sun's last glint; shining clouds which somehow form far above all others in the very coldest layer of our atmosphere. And we may here recall the poet Tagore's statement, "Birth is from the mystery of night to the still greater mystery of day." Is our familiar blue sky ultimately any less mysterious than the strange phenomena which you may never have heard of before reading this chapter?

One of the questions which small children traditionally ask is why the sky is blue. Like many of their questions, it is well worth our attention, if for no other reason than our desire to learn more about the beauty and apparent profundity of the clear sky of day. The blue sky is an object of inspiration, a fit subject for poetry and for our deepest enjoyment. But our aesthetic and spiritual appreciation of this incomparable sight can be enhanced by careful observation and scientific study. Most people would suppose it either difficult or impossible to find much detail to study in the plain face of a clear blue sky, but they are wrong, and an amateur observer is well equipped for this study with no more than his eyes and the very simplest and least expensive of materials. And the starting point for any observer is still that basic but profound question: why is the sky blue?

We know that the blue sky is not a solid vault at an infinite distance, although it gives us that beautiful appearance. The blue sky can only be in the atmosphere itself, and it exists only when that atmosphere is illuminated by sunlight (the blue effectively disappears at night). The white light which the sun sends us contains all of the colors of the rainbow. Could it be that water vapor and minute dust particles scatter the blue component of that light more strongly than they scatter the other wavelengths? The blue sky is certainly the result of preferential scattering, but not by these ingredients, because on humid, hazy days (when there is much moisture and dust in the air) we see *less* blue in the sky. The dust particles and water droplets must scatter all wavelengths about equally, so that on humid, stagnant days much white light is diffused and sent to us from all parts of the sky. The cloudless sky on such occasions shows only the slightest blue, which becomes altogether white in the vicinity of the sun (where the scattering of relatively large particles is especially great at low angles). On the vast majority of days, and especially later in the afternoon, there is a whitish area of considerable size about the sun. This area is called an *aureole* (the same term, unfortunately, is also applied to the innermost set of colors in a cloud-corona). There are different opinions about how the boundaries of this aureole should be defined; one way is to mark as its outer edge that distance from the sun at which the eye can first detect any trace of blue. Making sure *never* to look directly at the solar disk, measure (with your fist at arm's length, as explained on page 5) the angular radius of the aureole at the same time—or, rather, at the same solar altitude—each day. You will find (depending on where you live) that afternoons on which blue is visible right up to the solar disk are quite uncommon, but even more rare are those when the sky seems more blue than white right up to the sun's edge.

If dust and water vapor prevent the sky from looking blue, then there can be only one source of the beautiful color—the air molecules themselves! These molecules scatter the blue in white sunlight preferentially, so it is sunlight scattered by the very air we breathe which produces the blue vault of day. The sky of a clear day looks profound—profoundly deep for mental reflection—but once we understand the answer to the child's question, we can appreciate that the nature of the sky also *is* as profound as it could possibly be. Our atmosphere directly supports almost every life form on the planet, but life in turn helps to sustain this kind of atmosphere. The blue of the sky seems fragile and yet somehow also unvanquishable, but it is only the delight of it for those who experience it which is unvanquishable: the blue really *is* fragile, and we may read our own condition—and fate—in its hue. If the blue should disappear, the Earth as we have known it, the one which life and the blue sky have together made, will disappear, too. We would be left with the opposite of health, a grim environment for which we would find our senses ill adapted—including our senses of beauty and enjoyment of the natural world. The blue sky is for us not just an indicator or symbol, it really *is* health and home. No other planetary surface in the solar system has a blue sky, and this home world of ours, which has been called the Blue Planet for its oceans, could just as fittingly deserve that title for its skies.

What features can we observe and study in these precious heavens? Minnaert (p. 238) begins his treatment of the subject with a general description which could hardly be improved upon: "In unending beauty the blue sky spans the Earth. It is as if this blue were fathomless, as if its very depth were palpable. The variety of its tints is infinite; it changes from day to day, from one point of the sky to the other." Notice that the dreamlike grandeur in this description is nevertheless quite naturalistic: almost none of the terms are abstract in this context. How strange and wonderful that something which seems infinitely far (deep enough for deepest ponderings, high enough for highest aspirations) should seem "palpable"! But for specific study of the blue sky, Minnaert's most important reference here is to the variations in tint. As Minnaert remarks elsewhere (p. 247), John Ruskin (in his classic work, *Modern Painters*) "mentions the blue sky as being the finest example of a uniform gradation of colour." But what are the variations caused by, and how are they distributed?

Everyone has noticed that on some cloudless days the sky is bluer than on others, and we have already seen that the explanation lies in the amount of dust and water vapor in the air. But something which people seldom note—or would know how to explain—are the different blues in the sky at the same time. What is the bluest place in the sky? Three factors combine to give the location: they are the altitude of the spot above the horizon, the altitude of the sun, and the azimuth (horizontal measurement of position) of the sun. The higher in the sky we gaze, the less dust and water we look through, so we would expect the zenith to be the bluest point in the sky. The sun's position modifies the situation, however, because scattering from dust and water droplets is greatest at small angles (near the sun) and least 90° away from the sun. The bluest part of the sky can be determined with fair accuracy by simply observing and judging the hues, but the combination of the various factors gives us the following general rule: the bluest part of the sky is always on the vertical circle which passes through the sun (straight up from horizon below the sun to sun to zenith and to opposite horizon), with the point being about 65° from the solar disk when the sun is high, and about 95° when the sun is low. At sunrise or sunset the zenith is therefore the bluest part of the sky (a blue further enhanced by several other interesting factors).

Perhaps the most interesting endeavor in studying the blue sky is rating the deepest blue for a given elevation of the sun on different days. This can be done roughly with the exposure meter of a camera (the darker a part of the blue sky is, the bluer). There is, however, a far less expensive and generally superior way to rate the blue on different days. It can be done with one of the simplest of instruments imaginable, a cyanometer. A cyanometer is nothing more than a scale of blues painted on a card or on a series of cardboard strips. Add different amounts of blue paint to white paint in regular increments to achieve the various degrees of blueness, which can then be numbered. Each day when clouds do not interfere, go out when the sun is at the same elevation and see which of the colors on your scale most closely approximates the hue of the bluest parts of the sky. Be sure to let sunlight fall directly on your scale. A cyanometer, of course, can also be used to study the distribution of isocyans or isophotes (lines of equal blueness or lines of equal light intensity) across the sky at any given time.

Is it callous and cold to rate the blue sky numerically? It depends on how one uses the figures. The proper purpose of these ratings is to foster greater sensitivity to (and interest in) the more subtle beauties of the sky, and to help relate the deepness and distribution of blue to the context of weather in which it occurs. Put more simply, the ratings are not an end in themselves but a means by which one can become more aware and appreciative of the blue skies of all days. The average person hardly distinguishes between the blue of various clear days and is usually not consciously aware of those rare jewels of days when the sky is almost perfectly free of dust and water vapor and therefore a blue devastatingly deep and beautiful, more unfathomable and palpable than one would have thought possible.

The fairest and purest days are among the greatest joys any person can experience, so it is a triumph of living to be able to drink them in to the fullest possible. It is helpful to consider more closely here some of the many special features of the sky and landscape which can be enjoyed on such days, as well as the prior weather situation which may give rise to them. I choose one memorable example from my own experience. . . .

One might imagine that the very clearest days would occur after a long period of heavy rain has washed the atmosphere clean of almost all particulate matter. That generality seems to apply in the case of the clear day I am thinking about. Over 3½ inches of rain had fallen at all locations in a roughly statewide area two days before, to be followed by a day of overcast sky and drizzle. The cloud deck, whose back edge was sharply defined, was moving out after midnight revealing a marvelous scene. Stargazers have their own measures of atmospheric transparency, such as (on a given night) the faintest star visible to the naked eye and both the extent and amount of detail visible in the Milky Way (see Chapter Ten). One phenomenon which, when it occurs strongly, is not favorable for many kinds of astronomical observation is scintillation (twinkling) of stars caused by turbulence in our atmosphere. On this night of awe-some clearing, however, it was a stirring experience to watch the brilliant and seemingly innumerable stars leaping and flar-ing as madly as I have ever seen them, like hosts of silvery lamps or living spirits at some great festival in the fields of the heavens. I could already imagine the high-altitude winds that would be lowered to earth the next day, singing in the trees and whipping my hair (they had already stirred gale-force winds of clearing in my sense of wonder). And I eagerly pictured the dark blue of the sky which would greet me when I arose.

The following day the blue of the sky was deeper than I had been able to imagine it could be. Even people who normally would never look at the sky must surely have had difficulty in escaping the wonder of that blue, because it was dark down to such a low angular altitude in the sky that it was prominent even between trees and buildings which would be directly ahead in one's field of view. A still more remarkable feature of the sky down low was the compression of layers of different blueness into a relatively narrow band above the horizon. On this wonderfully clear day, even the visibility of nearby objects in the landscape was dramatically improved, and both shadow and sun were stronger and sharper. The sky was blue right up to the solar disk (no aureole). The few small fair-weather cu-mulus clouds which sailed by later in the day were astonish-ingly white (an alternative to rating the blue of the sky is judging the white of clouds at certain distances). And the bluest part of the sky? On a cyanometer I had made, I had set as the deepest blue a shade so dark that I wondered if I would ever see the sky approach that tint. But on this day the deepest blue in the sky (for my adopted standard elevation of the sun) was almost the equal of that extreme value on my scale.

Even all of these observations together do not define the

boundless beauty of a clear day. They are, however, exercises in attention to details; no full appreciation of something beautiful and interesting is without its special discipline. A labor of love for the area you live in is to see what are the days of bluest sky. It is also possible to take your cyanometer with you when you travel and to observe the blue skies of other regions or countries, or the sky from mountains. With a little practice, the observer becomes trained to judge the sky's blue even without reference to a cyanometer. He becomes like an artist whose eye and judgment are extremely sensitive to colors and their distribution—in the ever-changing painting we call the sky.

Besides the blue of the sky and the black, white, and gray of clouds, one of the most familiar of colors in the heavens is the red of the setting (or rising) sun. Why it is red is related to why the sky is blue. Since air readily scatters blue wavelengths of visible light, it stands to reason that a long pathway of air would scatter out almost all of the blue before it reached the observer. The longest such pathway is the one sunlight travels when the sun is low, a route which contains enough water vapor to absorb much yellow and green and, on hazy days, to scatter all other wavelengths except for one, the red.

Another strange feature of the low sun is its oblateness, its apparent compression in the vertical dimension. When the sun is on the horizon, this flattening is most extreme and the vertical diameter is only about five-sixths of the horizontal on an average day. This effect can be attributed to the fact that the greater thickness of atmosphere down low refracts light more strongly. The bottom of the sun is lower and therefore light from the bottom is refracted more, and raised in apparent position more, than light from the top. Refraction at the horizon is so strong that it normally raises the sun's apparent position more than 0.5°; therefore when we see the sun's bottom edge touch the horizon, the sun in reality has just gone below the horizon!

One of the most spectacular of all low-sun phenomena occurs as a result of the fact that, as we have seen, some wavelengths of light are refracted more strongly than others. Since blue and violet have been scattered and much yellow and orange absorbed by water vapor in the long air path down low, the light of the sun as it sets is predominantly red with, on clear days, a small component of green-blue. The additional scattering and absorption of hazy, humid days removes the green-blue also, leaving an especially red sun. But on very clear days the blue-green remains and, since this wavelength is refracted more strongly than red, there should be a green disk of the sun very slightly higher than the one of red light. If conditions are correct the very upper edge of the green disk should be visible just above the red one. When this occurs, the wondrous phenomenon that results at sunrise or sunset is called the *green flash.*

The green flash may occur in various degrees of intensity, each of which may give a somewhat different appearance. Names like the *green rim* and the *green segment* have been applied to the less dramatic versions to distinguish them from the green flash proper, but the terms may have contributed more confusion than organization to the study of this elusive, transitory phenomenon. The green rim is perhaps this sky effect when the green is only visible with optical aid as a slight tinge on the sun's upper edge, sometimes well before sunset. The green segment is supposedly more prominent and, according to Minnaert (p. 60), appears as a green tinge at the extremities of the partly set sun that spreads swiftly to the center, viewable with the naked eye for no more than about a second. Theoretical work shows, however, that the purely green section of the sun must always be extremely narrow—too narrow, in fact, to be resolved by the naked eye. Presumably then, the green segment is what is seen when the last sizable slice of the sun is *predominated* by green, whereas the green flash proper can only occur when the atmospheric conditions permit that last, unresolvably small all-green edge to be seen purely and distinctly, much like the brief appearance of a brilliant green star or line, which would look like a true flash.

The various names for the different appearances of the phenomenon have confused the issue of how frequently the green flash may be seen. I have certainly found many nights when conditions seemed proper, but not even with optical aid did the sunset show the slightest trace of a green rim. Alastair B.

Fraser, an expert in both the theory and the observation of the green flash, states that with optical instruments as modest as binoculars, the phenomenon can be observed as often as three or four times a month from most places which have an unobstructed view of the sunset. This figure no doubt applies to the observation of any degree of the green flash, so it is true that the green flash is not as rare as has often been maintained; occurrences that *are* rather rare are those more spectacular ones in which the phenomenon really does appear as a flash or even like an extended flame or ray. In some languages the green flash is called by the equivalent of the English *green ray*; this term is occasionally used in English as well.

But how could our simple mechanism explain the appearance of an actual beam or ray of light shooting up? There is no doubt about the authenticity of such reports. Authorities on the subject have been able to offer convincing models of unusual atmospheric conditions which could produce these wondrous appearances. It is even possible to have multiple green flashes! Greenler, in his book *Rainbows, Halos, and Glories*, gives a very clear explanation of how multiple flashes may occur. When a set of little spikes is seen sticking out of either side of the setting sun, the spikes may rise up to detach from the sun and disappear in flashes of green—the more spikes, the more flashes. Fraser, who first worked out the details of this situation, has written that his personal record is 21 green flashes at a single sunset!

It is even possible for the atmosphere to act like a magnifying lens. Various mirage conditions—extreme examples of refraction caused by temperature (and therefore density) differences—might cause different remarkable appearances of the green flash, even the sudden vertical extension which could suggest a ray. It was in fact as "the green ray" that the whole phenomenon was first brought to public attention, in turn stimulating the first serious scientific interest. The vehicle was Jules Verne's 1882 novel *Le Rayon Vert* ("the green ray," but perhaps the word ray here refers merely to the last light ray coming horizontally to the observer, not the spectacular vertical flame some observers have seen). A character in Verne's book relates an old Scottish legend that anyone who has seen the green flash will never again be deceived in matters of the heart, and will see clearly into the thoughts of himself and others. A romantic legend—for a romantic phenomenon.

The green flash normally is visible to the naked eye for no more than a second or two, but there have been some amazing exceptions in special situations. In lands of the midnight sun, it should be possible to observe the green flash for much longer as the sun creeps in its rising or setting, moving at a very shallow angle to the horizon. Members of Admiral Byrd's expedition to the South Pole were able to observe the green flash for 35 minutes. Sometimes an observer's own movement can prolong the flash: Minnaert (p. 59) watched it for 20 seconds by running up a slope, and another observer was able to see it several times in a row because of the motion of the ship he was on. The sun is not the only celestial object which can produce the green flash; on very rare occasions, it has been observed with the moon, with Venus, and even with Jupiter. The greatest prolongation of the green flash that has been observed occurred with an artificial light source: the phenomenon could always be observed, from a distant vantage point, in the light of a lighthouse. Minnaert also mentions a case in which someone could keep seeing the green flash by positioning himself at the edge of the shadow of a surprisingly close rock and then gazing over the rock (usually a distant horizon is required). One final kind of observation of the green flash must always remain fairly unusual: the observation of it at sunrise. This happens at least as often as the sunset green flash, but it is difficult to judge exactly when (and where) the first piece of the sun's upper edge will appear over the horizon.

The hue of green in this wonderful sky effect may vary considerably, but it is often very vivid. It is also possible to see a *red flash* when the lower rim of the sun peeks out from behind a cloud, but (as Greenler suggests) this effect is often overlooked because we are so accustomed to seeing a red sun. In his fine discussion of the green flash, Greenler also mentions that on the very clearest nights of all, a little blue survives the scattering and the green flash appears more blue than green. There have even been a few times when it has appeared distinctly violet!

Other less spectacular but more common effects can be observed on a low sun, especially with binoculars or telescope, but one must remember *never* to stare at the sun even with the naked eye unless the sun's intensity is very greatly diminished, as on a clear day occurs only in the last moments before sunset. With binoculars or telescope it is sometimes possible to see sunspots flattened out into rods and to admire especially well the fine gradation from orange to ever-deeper red between the top and the bottom of the solar disk. But always remember to exercise caution in observations of the sun: not even a glimpse of the green flash is worth risking damage to your vision.

There is at least one other low-sun phenomenon which is not extremely rare but may be as astonishing as the green flash. It is an instance of *inferior mirage*, so called because it is a mirage image which appears *beneath* the true image. This particular mirage of the sun occurs when there is a steep temperature, and therefore density, gradient not far above the surface. The light from a strip including the horizon is then refracted so strongly it is bent to pass far over the observer's head, never reaching his vision and therefore bringing closer together the true sun and its reflected image on a body of water. This particular effect has no widely accepted name of its own, but it certainly deserves one and I suggest we follow Minnaert (p. 55) in calling the image (and perhaps the whole phenomenon) the *countersun*.

The first time that I or any of my friends observed the countersun was over a stretch of about 12 miles of bay on an August day of average warmth but probably unusually high water temperature. Long before sunset we marveled at the sight of dramatic mirages. The one observation which I still cannot explain was the apparent vertical splitting of a distant ship into three parts, two large gaps separating the center section of the ship from its two ends! As we watched the sun's glittering path on the water finally shift into a reflection of the sun, we anticipated what promised to be a clear view of the sun all the way to the horizon. When the sun was still centered well over a degree above the apparent horizon, all of us burst into shouts of astonishment, for another sun (the reflected image) had suddenly lifted its upper edge above the distant verge of the world, and

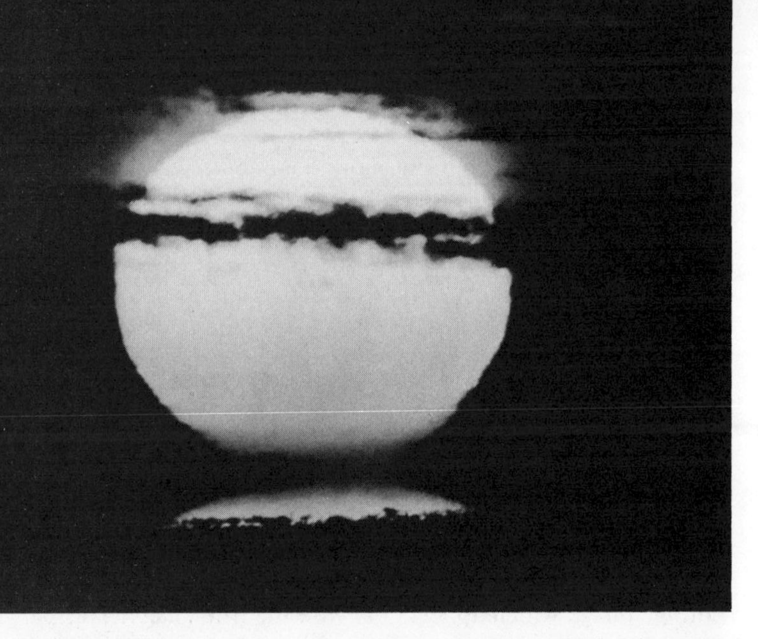

Fig. 13. *Sun rising from the sea, showing the gradation of color from yellow on top to red at bottom (see inside front cover). The top is partially obscured by clouds, while the bottom is reflected as a mirage image (the countersun) by a layer of warm air hugging the surface of the water. Photographed by George East on September 6, 1981, at Mashpee, Massachusetts, with a 1/1000-second exposure on Kodachrome 64 and a 1250-mm lens at f/10.*

was rising to meet the setting true sun! In most displays of this phenomenon, the countersun only manages to get a small fraction of itself up before it meets the true solar disk. But on this occasion, this wondrous image was able to raise far more of its bulk. I dropped my binoculars from my eyes because I suddenly realized that I also had to see this marvel with my unaided vision, partly just to convince myself that I could really be standing on a familiar beach in the real world with two perfect suns meeting in the sky before me. At its greatest height the countersun had climbed roughly one-half of its diameter; when it was thus half-risen, the two fiery globes touched and seemed to melt and start flowing into each other. The waist of light between them grew thicker, and now they looked like an angry mushroom cloud sitting on the waters. Finally, still blending, these awesome conjoined twin suns crept down ever lower until the last piece of fire was swallowed under the dark waves.

Although there are other distortions of the low sun which are not rare, let us move on now to consider the phenomena of the twilight sky. There is no more beautiful and mysterious time than twilight, which seems a whole world of its own or "the crack between the worlds," as it is called in the books of Carlos Castaneda. At no other time except storm's approach or total solar eclipse (which act as special forms of twilight) do we see such swift changes across the ever-beautiful face of nature. It is a period when the creatures of both day and night are still astir, and also when many of the sky phenomena of both day and night may be observed, sometimes in interaction with each other. Even better, of course, is the fact that in the sky there are phenomena which belong uniquely to twilight. Twilight is also commonly the only time when colors and differences in light intensity in the sky are so prominent that they draw the attention of almost everyone.

The colors which the uninitiated person usually notices are those striking ones which tinge clouds in the sunset and twilight sky. But the sky itself, from horizon to horizon, is washed over by waves of patterns and tints and varying intensities which, though more subtle, are also more intricate and more regular than the cloud colorings. The clouds are beautiful, but they are like impermanent islands in the overwhelmingly grander and more mysterious sea of twilight sky in whose system and embrace they live and lie. To use another analogy, the twilight phenomena are together like an orchestra which every clear evening plays the same complex and beautiful theme which would never cease to amaze even if it were performed identically each time. Instead, that same masterwork is set forth in endless variations of the theme. Minnaert provides a rich and lovely account of the standard development of the twilight phenomena, as well as of some of the major variations. In our present discussion, we will consider only a few of the most remarkable of the twilight phenomena. Their very names sound delightfully (almost preposterously) strange and mysterious: the purple light; crepuscular rays; the earth-shadow and Belt of Venus; volcanically enhanced twilights and ultracirrus clouds.

Of all these twilight phenomena, the *purple light* is the one whose *effect* is most conspicuous. Most people have noticed or imagined they noticed a strange increase in light intensity around the time when the sun has fallen to 4° or 5° below the

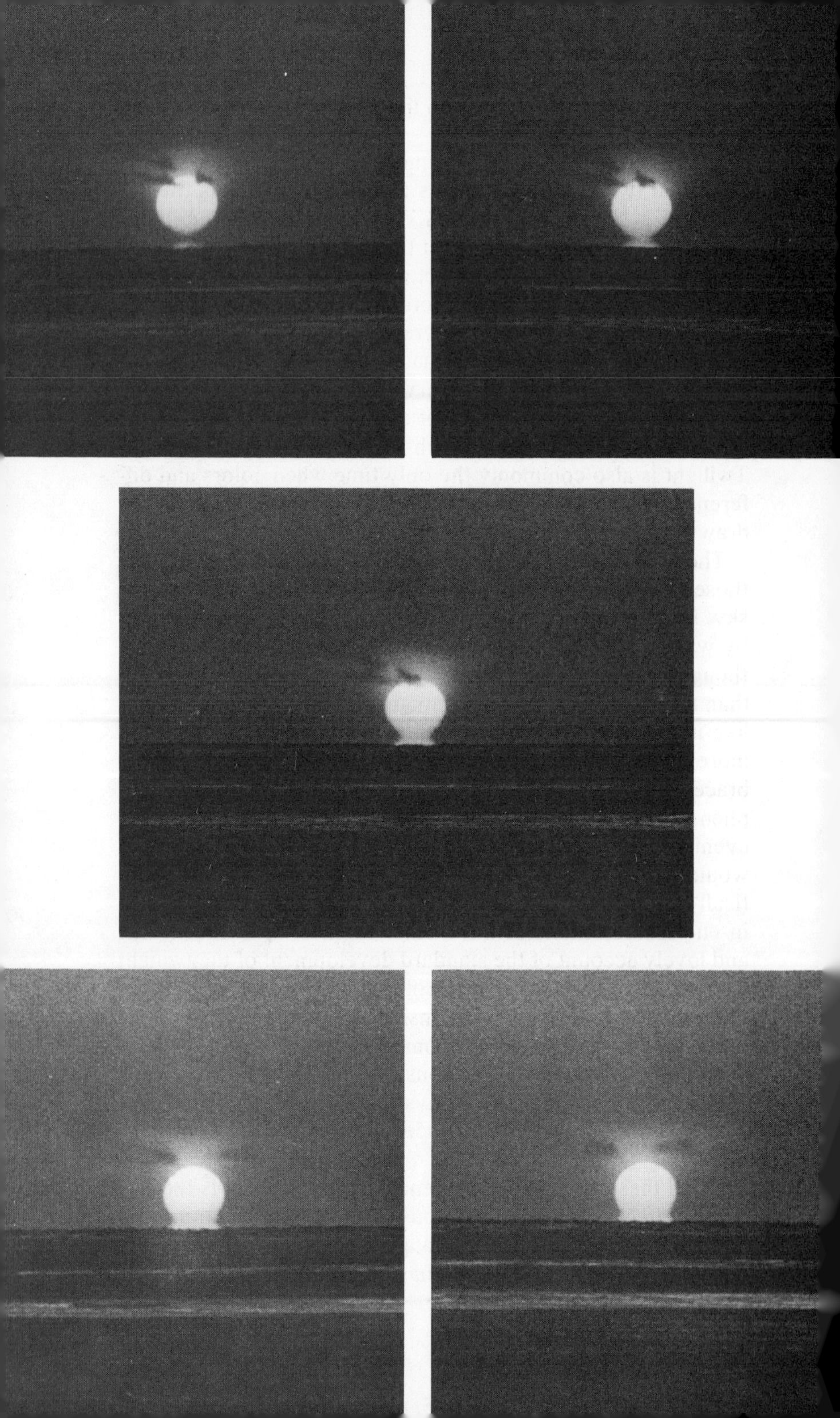

horizon (very roughly about 30 minutes after sunset depending on the observer's latitude and the time of year). Or one notices that around this time the western sides of all objects seem almost self-luminous with a warm pink or purplish glow. Could this light all be coming from the still-bright western sky? If so, why would the illumination seem for a while to get brighter? Could it be an optical illusion resulting from the fact that the eye requires time to adjust to the drastic drop in illumination which occurs when the sun disappears from the sky? The major factor is not the eye's adjustment, nor the light from all the western sky, but rather the light from a localized area which is surprisingly far above the western horizon. (One would naturally expect the brightest part of the western sky to be down low, nearest to where the sun set.) This is the area of the purple light.

The purple light, which is usually more pink than purple, first becomes visible roughly 15 to 20 minutes after sunset on clear nights. Measurements show that it is not actually a brightening, but an area which decreases in brightness less quickly than the surrounding sky. It is therefore not surprising that an observer often becomes aware of the purple light suddenly, without being able to describe the manner in which this large patch of diffuse light first seemed to form. Although the purple light is present on every evening, its intensity varies greatly. It may be more effectively and gorgeously distributed by thin layers of cirri, but the initial intensity itself is variable, and that fact brings us straight to the question of the purple light's cause.

The purple light is caused by scattering produced by a layer of very fine particles of dust which float at a height of about 12 miles in our atmosphere, or up to a much greater altitude after major eruptions. The light appears well above the horizon because it is the "secondary glow"—it is caused when the light which passes first through the layer enters into the layer again because that layer curves with the curve of the Earth itself. The purple hue sometimes observed when the light is extensive and long-lasting (after major eruptions) can be attributed to a

Fig. 14. *Sequence of photographs showing the rising of the countersun to meet the setting sun. In some cases the refraction is greater, and the countersun rises higher. Photographed by Chuck Fuller over Delaware Bay in January 1981.*

small component of the light which is scattered blue from the thin air still illuminated at high altitude (blue added to red gives purple). There can be little doubt that the particles are primarily the fine ash from volcanic eruptions around the world. It is noted that the purple light becomes extremely bright, long-lasting, and extensive (in the sky) over a large area of the world after a major eruption. Following the very powerful and ashy explosion of Krakatoa in 1883, the entire world was treated, for several years, to "colorful sunsets"—such is the phrase often used in books, and part of what is meant is bright, fiery, and colorful purple light (we will consider the other aspects of these volcanic sunsets in a moment). Since the stratosphere where the layer is located is for the most part calm, it is clear how the ash remains aloft for so long and how a certain amount can always be present to cause the purple light. Less clear is the nature of the material which causes the very rare occurrences of a second purple light, apparently visible even when the sun is almost 18° below the horizon. Whatever the material is (it is perhaps meteoric dust) it must be located at an altitude of about 45 to 55 miles above the surface.

Not all of the twilight phenomena are found in the western sky; with the setting of the sun there rises in the east the *earth-shadow* and, sometimes, the *Belt of Venus*. Many an outdoors person has noticed the blue-gray earth-shadow, widest at the point opposite the sun, without realizing that its top was the clearly visible edge of day and night, that this was the shadow of the Earth projected out through our atmosphere into space. For the first few minutes the earth-shadow ascends at about the same rate as the sun is descending (already below the horizon), but the shadow's climb begins to accelerate. The earth-shadow is always visible in clear, fairly unhazy skies, but the angular height to which its upper edge rises before becoming indistinguishable is an excellent measure of just how clear the atmosphere is at that time. Seldom is the top border distinct when it is more than about 15° above the horizon but occasionally it may be seen much higher.

The upper edge of the earth-shadow is sometimes bordered for a while by a band of lovely pink called (by some) the Belt of Venus. I do not know the origin or reason for this name, but it surely has nothing to do with the planet Venus (which could never be seen near it), so presumably the name refers to the goddess, who had a belt or girdle which could inspire in others passion for its wearer. A sight as romantic as the Belt of Venus is the full moon rising bronze-yellow from contrast with the earth-shadow which overtakes, surrounds, and passes it. A fascinating (but quite rare) opportunity is the chance to see the totally eclipsed moon rising or setting in the earth-shadow, for then we are seeing that shadow both on our atmosphere and a quarter million miles farther away on the moon's face. In bright twilight it might be quite difficult to observe the red hue of the shadow on the moon, but nonetheless that color may be considered a true stain of our twilight, or even an extension of the Belt of Venus (which forms a ring around Earth) into a solid moon-intercepted cone! For clarification of these lovely relationships, see Chapter Five.

The earth-shadow is beautiful—night seen from afar!—but also beautiful are the ultimate extensions of cloud-shadows and the illuminated breaks between them. In the midst of day we can often observe what is popularly called "the sun drawing water," or sunbeams or sunlight shafts of the sky; these are sunlight seen between cloud-shadows, made visible by very light mist. It seems strange that the rays coming from the distant sun, which must be virtually parallel to one another, appear to radiate like spokes from the sun, but this divergence is only an optical illusion such as we see when we look at parallel rails of a train track appearing to diverge from a point in the distance. These sky-beams of day sometimes have a cheerful effect, but they can also be eerily lovely, as is seen by their use in religious paintings. They are not so beautiful or so eerie, however, as their much less common twilight version, the *crepuscular rays*.

Crepuscular means "of twilight," and it is the backdrop of the twilight sky—especially the purple light—which helps make the crepuscular rays so prominent. The rays still appear to radiate from the sun, which is now below the horizon, but they form a vast fan of stripes of pink sky and shadow. The contrast of both colors and light intensities is more favorable, and now the shadows are far longer, extending for at least

Fig. 15. *Crepuscular rays herald a new day over the North Dakota prairie (see inside front cover).*

hundreds of miles: the clouds which cause crepuscular rays are usually so far to the west (or, in morning, the east) that they are below the horizon! Crepuscular rays can therefore give us some general information about weather that is perhaps several states and maybe an entire day away.

This phenomenon can sometimes be observed diverging from the antisolar point just above the eastern horizon as *anticrepuscular rays*, each one matching a crepuscular ray in the west. On rather rare occasions it is possible to see the rays extending all the way across the sky, with the effect of perspective making them appear curved.

The awakening of Mount St. Helens in 1980 provided American observers with some superb displays of volcanic effects on twilight. The average person in most of the continent was not enough attuned to the details of twilight to notice more than offhandedly some of the spectacular twilights, and generally did not associate them with the volcano, but Mount St. Helens tinged every life a little with painted evenings. Then, in 1982, the El Chichon volcano of Mexico released an ash-cloud perhaps ten times greater and initiated months of twilights probably more vivid than any since at least 1912. The most prominent feature across most of America in both the 1980 episodes and throughout 1982 was an intensification and enlargement of the band of color which appears low in the west for a while after sunset on clear nights. This band is atmosphere to the west of the observer which is still strongly illuminated and scattering light from the sun (it is the "primary glow"; the purple light is the secondary). Normally it is confined to within a few degrees of the horizon; it is only distinct on rather clear days and during a relatively short period a while after sunset, when it is typically orange. What a change is wrought by the presence of large quantities of volcanic ash! Sometimes the band becomes visible well before sunset, in which case it is white or gold. The most prominent time for the display is still roughly 15 to 40 minutes after sundown, when the band is

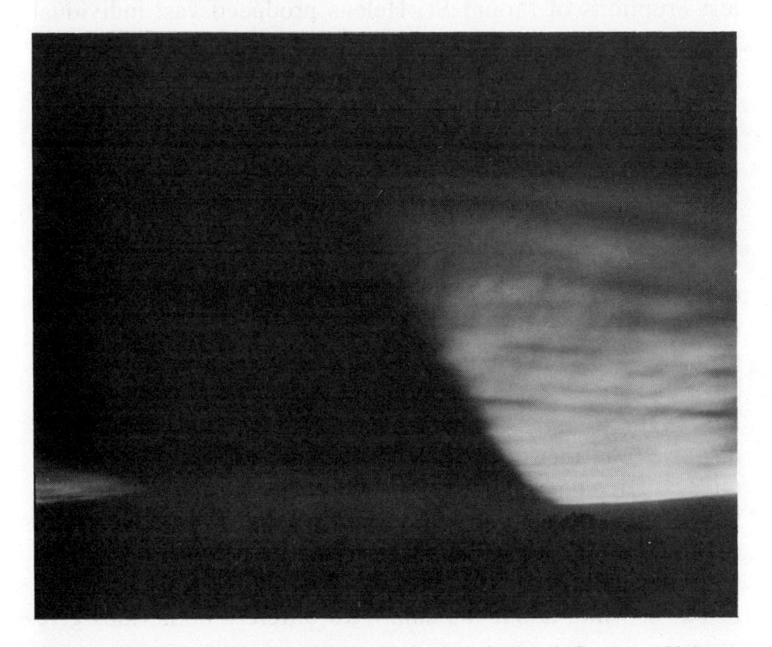

Fig. 16. *The ash-cloud of the Mt. St. Helens explosion drifts over Oklahoma City, showing crepuscular rays. These were caused by the shadow of cumulonimbus clouds beyond the horizon in Texas.*

orange, but in volcanic twilights the band can be visible to several times higher in the sky and its intensity is sometimes so strong as to be incredible. A discrete zone in the western sky seems literally painted with burning color that changes from orange to red and sometimes finally even purple as it slowly descends to the horizon. This is primarily what is meant by the "colorful sunsets" caused by volcanic ash!

After major eruptions, ash can also be seen in greater concentrations forming thin clouds. These are what Minnaert calls *ultracirrus clouds*, indicating that they are higher than typical cirri, yet perhaps seldom more than 7 miles up. They are much too thin and delicate to be seen during the day, so it is only for a while after sunset (rarely, also a while *before* sunset) that they become visible. The exact height they reach (both in the sky and in the atmosphere) depends on such factors as the eruption, the atmospheric circulation, and the observer's geographical position.

Although these thin gray (or slightly red-brown) ultracirri are not dense enough to be noted in broad daylight, the mightiest eruptions of Mount St. Helens produced vast individual ash-clouds which in at least one case actually dimmed the daylight on the East Coast three thousand miles away. The morning darkness and ashfall which these great clouds caused across the Pacific Northwest are well known. But known to only a few people are the very interesting effects of the second of these clouds on the skies of the eastern United States. The first cloud—from the eruption of May 18, 1980—crossed the eastern seaboard at night when much of the area was under low rainclouds. But the cloud from the May 25 blast came over at a height of about thirty thousand feet, and for several hours dramatically decreased the illumination from an otherwise clear, haze-free sunny sky. From a location near the East Coast later in the summer and autumn, I was able to observe distinct displays of ultracirri, but this heavier cloud may also have produced an even more interesting effect: I believe that I detected faintly the seldom-reported *Bishop's Ring*. This is a corona caused by thick ash-clouds whose tiny particles diffract light, producing a huge red or red-brown ring that surrounds an area of bluish white about the sun. Two different authorities give the average radius of the reddish circle as 15° and 28°—far larger than the ordinary corona caused by water or ice.

Bishop's Ring was first documented after the stupendous explosion of Krakatoa in 1883, when it was seen by the Reverend Sereno Bishop in Honolulu and soon after by many other observers around the world. What I saw in 1980 was certainly no ordinary aureole of the sun. The bluish white area was ringed by a very indistinct brownish edge and the radius was about 15°. Observers in some parts of the United States should have been able to see a rather prominent Bishop's Ring, but once again we are reminded that it is difficult to notice even striking sights for which we do not already have a place in our intellectual framework. I may also have seen Bishop's Ring caused by El Chichon; more southerly observers did report some sightings in 1982. It is certainly quite possible that Mount St. Helens will have more ash-producing eruptions in upcoming years, so skywatchers throughout the United States may have some excellent continued opportunities. How remarkable that such terrifying violence in nature can give rise to some of the most delicate and pretty phenomena!

Delicate and lovely are words that well describe the sight of a slender crescent moon afloat on the shores of the sun's afterglow, set on the edge of the deepening shades of night. This is a *young moon*. (The age of the moon is the amount of time that has elapsed since the last occurrence of the *new moon*, the invisible phase which occurs when the moon passes due north or south of the sun.)

When a young moon is more than about 28 hours old it may appear in a sky dark enough to permit sight of *earthshine*, the usually very dim illumination on the dark, unsunlit part of the moon's disk (this phenomenon has long been known popularly as "the old moon in the new moon's arms"). Earthshine is just what the name suggests: it is light from our planet's daylit side falling on part of the moon's night side. Once the moon has waxed to become a very thick crescent the earthshine is difficult to observe—not only because it is overwhelmed by the sunlit lunar surface (the bright part of the moon), but also because there is then a thinner piece of sunlit Earth shining in

Fig. 17. *The evening crescent moon on January 8, 1981, was strongly illuminated by earthshine. This unusually clear picture of earthshine was taken by Kerry Hurd of Medfield, Massachusetts, at the Newtonian focus of a 10-inch, f/6 reflector, with a five-second exposure on Fujichrome 400.*

the skies of the moon (the phase of Earth seen from the moon is always the complement of the lunar phase visible from Earth). On quite rare occasions, earthshine may be distinctly visible to the naked eye even after first quarter. Why, however, should its intensity vary from one month to the next at the same phase? Because the daylit Earth shining on it may be more or less reflective according to how much water, land, and cloud are in view (clouds are most reflective, producing brighter earthshine). Earthshine therefore shows us general weather conditions in a distant part of our planet, and it is very interesting to rate its intensity each month for each day of the crescent phase.

If, however, you want to see the greatest kind of marvel of all—a very common thing seen in a most uncommon state—then you can undertake the quest for the youngest moon. Most people have at some time or other spotted a crescent of 30 or 35 hours, and been amazed at its slenderness, charmed by its loveliness. But an extremely young moon, one of 24 hours or less, is a sight so strange it is truly breathtaking. The quest to observe the youngest moon you can is a lifelong one, and the young moons you get to see are objects as enchanting, beautiful, and strange as any jewel or magic talisman in a fairy story. The youngest crescent I have yet observed was one 24½ hours old. That wire-thin arc of light was shortened at its ends. (The shadows of lunar mountains begin to eat up the ends of the crescent so that there is a limit to the youngest moon which can be seen at all. The critical measure which determines roughly what will be seen or not seen is the moon's angular distance from the sun. The astronomer A. Danjon surveyed successful and unsuccessful attempts to observe very young moons and concluded that the last piece of crescent disappears when the moon comes to within about 7° of the sun.)

The 24½-hour-old moon I observed was truly unlike anything else I had ever seen. It gave me a sensation of wonderful strangeness, especially because it made me realize once again that the familiar observing site in my neighborhood was not so ordinary after all—not if such a sight as this could hang over it. It was thin as a single snippet of pale gold hair falling, falling gently through dusk to the repose of the low forest horizon. Or perhaps it was a slight but dreaming and luminous smile scarcely touching the face of that twilight sky, a face whose tender tones and shades were the only things which could possibly be delicate enough to hold that moon. Fragile? Yes, the crescent seemed exquisitely so, but at the same time I had the feeling that this was an indelible mark, a slenderest sliver of pure celestial beauty that eternity keeps forever from harm or slightest alteration. I knew at least as it finally slipped into the utter darkness that this moon was a craftsman's finest slice in my mind, from which the sparks of imagination would long fly—they are flying still.

This observation of mine was by no means a record for the youngest moon. Not too many people have seen younger ones, but extremely thin crescents are not difficult to glimpse if astronomical conditions are proper and weather is good. In general, late winter and early spring are best for seeing a young moon because the moon's path in the sky is then most nearly vertical (for mid-northern-latitude observers). For any given separation of the crescent from the sun, the moon is then higher above the horizon. An astronomical almanac which for each year takes into account this and other factors to plot the position of young moons is Guy Ottewell's *Astronomical Calendar*. This guide also provides diagrams for all of the year's *old moons* (the old moon is the lunar crescent seen as soon as possible *before* new moon). Many general almanacs provide at least one important piece of information: the precise time of new moon. This enables one to calculate exactly how old the moon is on particular evenings. (Thus you can know that it would be impossible to see a young moon at sunset if new moon had occurred just a few hours earlier.) If you are aware of the various young-moon possibilities for the year, and you really do go out to observe, seeing a very young moon is simply a matter of waiting for one of those evenings to be reasonably clear.

Seeing young moons does not necessarily take trained observational skills. Proof of this is the fact that some extremely narrow crescents have been spotted, sometimes accidentally, by people with no experience at all as amateur astronomers. As a matter of fact, the all-time record for the youngest moon ever seen with the naked eye was a 14½-hour-old crescent seen accidentally by two housemaids, Lizzie King and Nellie Collinson, on a very clear night in England, May 2, 1916. It is almost certain that this observation was genuine because another person in England independently saw the crescent about 15 minutes later, and none of these observers could have confused the date—for this was the very night of a Zeppelin raid over Yorkshire. How shameful that a night of such memorable and miraculous celestial beauty was stricken with the terror and destruction of war. We must be more sensitive to this kind of beauty, because it makes us more sensitive to the beauty which we can and should find in ourselves.

A very young moon usually sets no later than about 45 minutes after sundown, but this is late enough for it to shine with considerable luster in a fairly well darkened sky. But is it still a twilight sky? The question of when twilight ends depends ultimately on how far the sun is below the horizon. *Civil twilight* ends when typical outdoor activities first become difficult without artificial lighting; *nautical twilight* ends when there is not enough light for the sea horizon to be visible; *astronomical twilight* ends when the last trace of the afterglow disappears in the west. It is debatable exactly how far the sun must be below the horizon to create these effects, but the traditional figures are solar altitudes of $-6°$ for civil twilight, $-12°$ for nautical twilight, and $-18°$ for astronomical twilight. In his article "Some Thoughts on Twilight" (*Sky and Telescope* magazine, October, 1960), the late Joseph Ashbrook provides some fascinating background on how these definitions were decided and discusses related matters. One of the most interesting twilight events he mentions is the passage of the *crepuscular arc* through the zenith. This arc is the line where the light-intensity gradient is steepest between the bright western and the dark eastern sky.

The definition of twilight is relevant to the understanding of the next phenomenon we come to in our journey, a phenomenon higher above the surface of the earth than any we have yet encountered save for the moon (earlier, we were considering what were primarily effects of the terrestrial atmosphere on the appearance of the sun, not the sun itself). In the last chapter we explored the lovely, high-floating nacreous clouds; now we come to the one variety of cloud which is found at a far greater altitude in the atmosphere. Whereas the nacreous can glow for several hours after sunset, *noctilucent clouds* can be illuminated all night long. The very name is Latin for "night-shining," and is sometimes translated from European languages into English as "luminous night clouds."

Strictly speaking, however, the name of these clouds is inaccurate, for neither nacreous nor noctilucent clouds are illuminated during the period of true night—that is, in the time between the end of astronomical twilight in evening and its beginning in the morning. Both of these cloud types are by far most often visible at high latitudes (no noctilucent clouds in the Northern Hemisphere have ever been seen as far south as New York City), and, in extreme northern and southern lands, noctilucent clouds occur in summer, when there is no real night. Noctilucent clouds are, in fact, exclusively a phenom-

TABLE 1

Table of Increasing Depth into Twilight

Solar Altitude (degrees)	Minutes After Sunset[1]		Illumination of a Horizontal Plane[2]
+.60	−3	Zero-magnitude stars (Vega, Arcturus, Capella) visible with naked eye if one knows just where to look.[3] (Also confirmed by present-day observers.)	.
0	0	Green flash proper.	
−1	+5	Magnitude 0.7 or 0.6 object (Saturn on Feb. 20, 1982) visible with naked eye if one knows just where to look (−1.1° solar elevation).[4]	250
−2	+10		113
−3	+15	Star Regulus (magnitude 1.3) visible to naked eye if one knows just where to look.[4]	40
−4	+20	Purple light brightest (−4 to −5° solar elevation).	13
		Second-magnitude stars visible (−4.3°).[3]	
		J. Ashbrook reported that Polaris (magnitude 2.0), the North Star, was first visible to him at solar alt. −4.8° on average, for another observer just −4.2° (see Aug. 1969 *Sky and Telescope* magazine; try this test for yourself).	
−5	+25	Purple light disappears (−5 to −6°).	5
		Third-magnitude stars visible (−5.3°).[3]	
−6	+30	Civil twilight ends.	2
		Crepuscular arc passes overhead (−6.4°).[5]	
		Fourth-magnitude stars visible (−6.8°).[3]	

TABLE 1—*Continued*

Solar Altitude (degrees)	Minutes After Sunset[1]		Illumination of a Horizontal Plane[2]
−7	+35	Reading outdoors no longer possible (−7.2°—J. Ashbrook tested with print of *Sky and Telescope*).	
−8	+40	Fifth-magnitude stars visible (−8.9°).[3]	
−9	+45		
−10	+50		
−11	+55	Sixth-magnitude stars visible (−11.6°).[3]	
−12	+60	Nautical twilight ends.	
−13	+65		
−14	+70		
−15	+75		
−16	+80	Last lighting of noctilucent clouds (about 50 miles high); end of the rare second purple light.	
−17	+85		
−18	+90	Astronomical twilight ends (last trace of afterglow gone).	

[1] Precise figures will vary according to latitude and time of year; these are for around 40°N at about the time of the equinoxes.

[2] With a cloudless sky, of course. Figures in light-intensity units called lux; derived from Minnaert.

[3] Figures from nineteenth-century astronomer J. F. J. Schmidt.

[4] Author's observation.

[5] From a series of observations by J. Ashbrook.

enon of deep twilight, for they are normally too faint to be identified if the sun is less than 6° below the horizon and cease to be illuminated when the solar depression is about 16°—before the end of astronomical twilight. Even during this time they can usually only be spotted in the region of sky just above the bright band of primary scattering which indicates the departed sun's position below the horizon.

Noctilucent clouds are not just curiosities; they can be among the most splendid sights in the heavens. There is an old saying that every cloud has a silver lining, but noctilucent clouds shine *entirely* silver-blue—except that they sometimes have a golden lower edge! One might expect that such high-altitude objects would not show signs of turbulence, but noctilucent clouds may come in waves, "billows," and "whirls," and they have been observed moving at speeds ranging as high as 100 to 500 miles per hour. The altitude of these clouds is always found to be almost precisely 50 miles, roughly two-and-a-half to four times as high as nacreous clouds, and at the same altitude as the lowest auroras. This is also precisely the altitude at which many meteors disappear, and there can be little doubt that noctilucent clouds are at least partly composed of meteoric dust, apparently coated with some amount of ice.

Striking proof that some noctilucent clouds are largely meteoric dust was provided in 1908 when a small piece of a comet exploded over Siberia and afterwards produced a breathtakingly strong and widespread display of noctilucent clouds all the way to Britain (comets carry vast quantities of meteor dust with them—see Chapters Four and Five). But could noctilucent clouds possibly owe part of their composition to volcanic dust shot up from the most violent eruptions? They occur in the area in which, it has been proposed, floats the dust that creates the rare second purple light. Is there a connection between major eruptions and occurrences of both second purple light and noctilucent clouds?

It is interesting that nacreous and noctilucent clouds form at the altitudes in our atmosphere where temperatures are lowest. The temperature falls as we go up from the surface to the bottom of the stratosphere, where nacreous clouds are, then rises (back up to about 32° F) at the top of that zone before decreasing to a deeper minimum at the top of the mesosphere, where noctilucent clouds are found. (At still higher altitudes the temperature increases greatly, but we must remember that in these regions of tenuous atmosphere, temperature has meaning only as a measure of the speed of molecular motions.)

One of the best places to observe noctilucent clouds is Alaska in the summer, though at that time the sky there is unfortunately often obscured by lower clouds. Another is Scandinavia, where the noctilucent clouds have even been glimpsed as high as the zenith on certain occasions. It has long been stated in books that noctilucent clouds have never been observed in the Northern Hemisphere south of 45°N. (This still puts them within the range of the northern continental United States and certainly Great Britain and a large part of Europe.) But recently, Steve Albers, after sighting some fairly conspicuous noctilucent clouds from North Dakota, recognized that he had earlier made an observation of what were probably noctilucent clouds from near Deposit, New York—just 42°N. Nowadays it is possible (indeed almost unavoidable) to see many clouds illuminated by distant towns and cities, so there is a possibility of confusing these with dim noctilucent clouds. Despite that difficulty, the Albers observations make it clear that anyone who lives at a fairly high northern latitude should check the sky low in the north during the deeper part of the long summer twilights. It would be a pity to miss a chance to see the splendid noctilucent clouds!

Low in the north sky is also the place for many United States viewers to look, at any time of year, for the beginnings of an *aurora*, which in the northern hemisphere is commonly called the Northern Lights. Auroras are quite well known to those fortunate people living around the oval of maximum auroral activity located about 23° away from the magnetic north pole in northern Canada. They are most frequent and strong during, and possibly for several years after, a maximum in the roughly 11-year cycle of solar activity. It has long been recognized that events on the sun can be correlated with magnetic storms on the Earth; the initial disturbance of the Earth's magnetic field begins as soon as a flare or a large sunspot group is pointed towards us, but auroras commence about two or three days later. The endless stream of atomic particles called the *solar wind* brings these particles to the environment of the Earth, but our planet is protected for the most part by our magnetic field, which is deformed by the solar wind into a kind of wake called the *magnetosphere*. On the side of the Earth away from the sun, the magnetosphere narrows into a magnetotail in which low-energy electrons from the sun are increased in energy by the disturbances in the magnetic field (magnetic energy is, in some manner not well understood, transformed into kinetic energy). These electrons follow magnetic lines down to the atmosphere in the regions of the magnetic poles, where they collide with atoms and molecules in the upper levels of atmosphere. This excites and ionizes the atoms and molecules, producing the eerie fluctuating patterns of light we call auroras.

So much can be said about auroras that we must be very

limited in our discussion here. The geographic ovals in which auroras primarily occur are about 23° from the magnetic poles, but Earth's rotation under them increases the favored areas of observation into bands. Since the Northern Lights can extend from an altitude of as low as about 50 miles to one of as high as about 600 they can often be seen low in the northern sky by much more southerly observers, even during rather typical displays. In unusually strong auroral outbreaks the Northern Lights may be rather prominent as far south as roughly 35°N in the Western Hemisphere, and they have supposedly been seen as far south as Singapore, which is 1½° north of the Earth's geographical equator, but south of the magnetic equator.

The terminology used to describe the many varieties of auroral patterns and motions is perhaps more extensive than it needs to be for either efficiency or clear aesthetic appreciation, but much of it certainly is beautiful, with expressions like *quiet rays* and *draperies*. Although there is certainly considerable complexity in what is seen, much of this results from the different geographic positions of observers or from combinations of effects: rays join to form *curtains* and a section of curtain seen overhead may appear as a zenithal crown of rotating, pulsing beams called an *auroral corona* (quite different from cloud-coronas or the corona of the sun). The best time to look for an aurora is at the midpoint of the night. A fine auroral display may begin with an amorphous auroral glow low in the north, which afterwards gives way to one or more horizontal arcs. These arcs sometimes rise one after another and fluctuate in brightness along various parts of their length; for more southerly observers this may be the most impressive phase of a display. But if one is fortunate—or northerly—vertical rays form on the arcs, combine to make curtains rippling along their "folds," and extend far over into the southern sky, perhaps even uniting in that ultimate color-organ, the auroral corona. The different possibilities of pulsings, flickerings, undulations, and shooting spears, in all speeds and combinations, are countless. In their motions these symphonies of light may be very slow and serene or startlingly restless and active, as various and original as their musical counterparts.

The vividness of fine auroras also staggers the eye. In many

Fig. 18. *An auroral drapery rapidly ripples as the rays paralleling the Earth's magnetic field shift about.*

Table 2

Table of Increasing Height into the Atmosphere

Layer		Altitude (Miles)	
troposphere	sea level	0	
troposphere	Mt. Everest	$5\frac{1}{2}$	
troposphere	ash-cloud from Mt. St. Helens causing possible Bishop's Ring	$5\frac{1}{2}$	
troposphere	high cirri	7	
troposphere	ultracirri (volcanic ash)	7	(and up?)
troposphere	tropopause (top of troposphere)	varies 5–10	
stratosphere	tops of tallest thunderstorms	12	(and up?)
stratosphere	volcanic ash causing ordinary purple light	12	(and up?)
stratosphere	greatest quantity of ozone	12–13	
stratosphere	nacreous (mother-of-pearl) clouds	12–20	
stratosphere	first minimum of temperature, about −60° F (after this, temp. rises again)	12–20	
stratosphere	highest airplane flight for plane leaving ground under own power	20	
stratosphere	relative maximum of ozone (ozone forms highest percentage of the air that it does at any altitude)	20–25	
stratosphere	highest balloon flights	25	
stratosphere	low meteors disappear	25	
stratosphere	maximum effect of ozone heating (temp. often as high as 32° F, highest ever was 86° F)	32	
stratosphere	atmospheric pressure 0.001 of sea level	32	
stratosphere	stratopause (top of stratosphere)	32	
mesosphere	height of volcanic ash from Mt. Agung causing a particular prolonged purple light in 1963	37	
mesosphere	atmospheric pressure 0.0001 of sea level	40	

Sphere		Altitude (Miles)
mesosphere	noctilucent clouds	50
mesosphere	second minimum of temperature, about −130° F	50
mesosphere	layer causing rare second purple light	50
mesosphere	lowest edge of auroras	50
mesosphere	low-burning meteors begin to shine	50
mesosphere	mesopause (top of mesosphere)	50
thermosphere	lowest visible airglow	50–60
thermosphere	high-burning meteors disappear	60
thermosphere	homopause (top of homosphere, bottom of heterosphere—all the atmosphere can be divided into these two sections independently of the other spheres; above the homopause air is no longer a fairly homogeneous mixture of its constituent gases)	60
thermosphere	atmospheric pressure 0.000001 of sea-level	60
thermosphere	flights by X-15 rocket plane	65–70
thermosphere	atmospheric pressure 0.000000001 of sea-level (temp. back up to 100° F—but only meaningful as a measure of molecular and atomic motion)	70
thermosphere	high-burning meteors begin to shine	80
thermosphere	practical lower limit of space (lower orbit is not possible for prolonged free flight, due to atmospheric drag)	100
thermosphere	atmospheric pressure 0.000000000001 of sea-level	100
thermosphere	thermopause (top of thermosphere)	varies 175–250
exosphere	average height of exobase (bottom of exosphere)	300
exosphere	highest airglow	300–400
exosphere	upper edge of highest aurora	600

NOTE: Figures are approximate in many cases.

cases an aurora can easily overwhelm the brightest stars. The aurora has even been seen in broad daylight! The most common auroral color is pale green or greenish yellow, caused by the electrons' effect on oxygen atoms at lower altitudes; red is caused by hydrogen and oxygen at higher altitudes. In addition to these basic colors, a more vivid green, purple-red, orange, and both blue and violet are sometimes observed.

There was a famous sighting of the Northern Lights in the days of the Roman Empire, and the phenomenon must surely have been observed on rare occasions in the other great (but not very northerly) civilizations of ancient times. Many of the cultures of northern Europe and Australia made the aurora into ghastly or frightening parts of their legends: the campfire of a demon, the spirits of the murdered, or a vortex of fire which swallows all the rivers of the world (Rutja's Rapids in Finnish myth). The Norse revered the lights as flashings from the noble Valkyries. Many of the native American peoples saw them as shy and pretty dancers, sometimes the spirits of the dead dancing in joy. Seemingly in the realm of lore, but possibly grounded in fact, are modern reports of sounds from auroras. It is difficult to imagine how sounds could be propagated all the way to the surface from the thin gases so high above, but observers claim that these noises corresponded to visual auroral activity. They were heard as cracklings or swishings (like those of a silk dress!). There certainly remains much to explain in this most various and mobile of all colored phenomena in the sky.

If auroral flames and beams do not dance from dark to dawn, we come at last, beyond twilight, to the deep dark of a moonless night. Is such a night, however, really ever a truly black background for the several thousand stars visible to the naked eye in the country? No. After one's eyes become adjusted to the low level of illumination, the sky is seen to be gently luminous, a beautiful gray behind the black silhouettes of trees and other objects in the landscape. That gray sky is direct proof that far more than stars visible to the naked eye enables us to find our way easily on a clear night. What is this light of the night sky itself? Its three major components are, on the average, of roughly equal proportions. The *airglow* (a kind of general permanent aurora—distinct from the polar aurora—that is caused probably by several factors and occurs at a height of more than 50 miles) and the combined light of telescopic stars (i.e., stars visible only by telescope), brightest in the Milky Way band, are two of these components.

But it is the last which is probably most interesting to observe: the final constituent of the night sky's brightness is the sunlight which is scattered to us by meteoric dust far beyond Earth's atmosphere, dust that exists throughout the solar system and is constantly replenished by comets. Most of this material is concentrated in the plane in which the Earth orbits, whose projection in the sky is called the *ecliptic*, the central line of the zodiac. As twilight draws to a close on a clear night, one may therefore see a more or less pyramidal form of diffuse light extending up from the afterglow along the ecliptic, and consequently referred to as the *zodiacal light*.

In the country the zodiacal light is sometimes prominent enough to be noticed by a casual observer, but even then it is often mistakenly thought to be the last part of the sun's after-

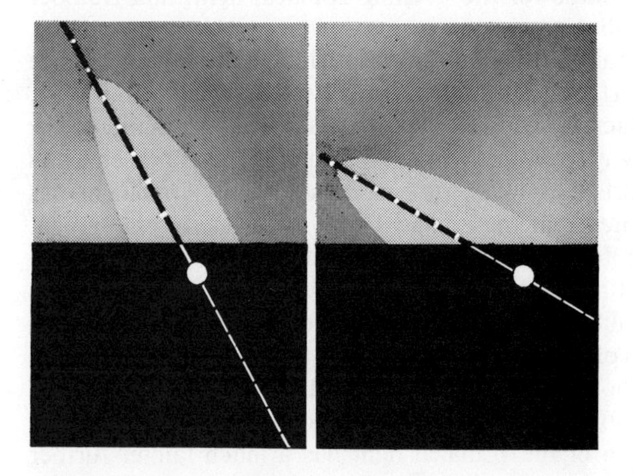

Fig. 19. *Position of the evening zodiacal light around spring equinox (left) and autumn equinox (right) for north temperate latitudes. The positioning around the solstices would be intermediate between these two extremes. Dotted line represents the ecliptic.*

glow in our atmosphere. Before the advent of the ubiquitous artificial lighting in this century, however, there were careful skywatchers who noted the zodiacal light as a distinct phenomenon. In his famous *Rubaiyat*, the twelfth-century Persian astronomer, mathematician, and poet Omar Khayyam mentions the "false dawn," the appearance of the zodiacal light in the eastern sky before the beginning of morning twilight. One of the first appraisals of the zodiacal light by a scientific mind was at the very end of the eighteenth century, when it was studied in the tropics of South America by the great naturalist, explorer, and observer Alexander von Humboldt. It was not only the climate of the tropics which aided von Humboldt's observations, but also the steeper inclination of the zodiacal light to the horizon at all times of year in those equatorial latitudes. At the higher latitudes of the temperate zones, the pyramid of light lies at a very shallow angle to the horizon around the times of the solstices, so it is then lost in the low mists and last vestiges of the afterglow. Only for a few months around the equinoxes is the angle very favorable for middle latitudes. In the Northern Hemisphere, February and March are good months for the evening zodiacal light, and October and November for the morning version.

The central portion of the zodiacal light may sometimes have a greater surface brightness than even the brightest Milky Way, and is in fact more milky, less granular (the "grains" of the Milky Way are, of course, stars). Amateur astronomer Chris Schur reports that under the best conditions in Arizona he has visually noted that the zodiacal light is distinctly blue-green, and his photographs prove it (for a discussion and examples of Schur's fine all-sky photos, and how he takes them with the use of an automobile hubcap, see the June 1982 and August 1982 issues of *Sky and Telescope*). On favorable nights far from the "light pollution" of cities, the tall pyramid can extend for 30° or 40° along the zodiac at the end of twilight; in reality this main section of the zodiacal light has a much fainter further extension which arcs all the way across the sky to reach the top of the morning zodiacal light. This very dim bridge between the pyramids of evening and morning is called the *zodiacal band*. The band is of about the same brightness along much of its length, but intensifies slightly at the antisolar point to form the *gegenschein*, or in English the *counterglow*. The total brightness of the counterglow is comparable to that of Sirius, the brightest star, but this brightness is distributed over an area at least three or four times wider than that of the moon. It is therefore necessary that one look for the gegenschein on moonless nights, and when its position is not covered by the Milky Way. Yet this roughly elliptical patch of light is not as difficult to observe as is often suggested. I have seen the gegenschein and the still fainter zodiacal band rather prominently on several occasions, though one usually needs to focus one's gaze slightly to the side of them in order to use the more sensitive areas of the retina. One observer I know tells me that on some nights in the Pocono Mountains of northeastern Pennsylvania (a few hours' drive from New York City) he has seen the gegenschein when it was ''spectacular''! The discovery of this counterglow and the zodiacal band are often attributed to T. J. Brorsen but the former was mentioned far earlier by von Humboldt and perhaps reported first of all by Esprit Pézénas in 1731.

The zodiacal light, band, and counterglow are of course all parts of the same phenomenon. The gegenschein is caused by dust outside the earth's orbit backscattering light like countless tiny full moons, making the gegenschein slightly brighter than the rest of the zodiacal band. The zodiacal light and its extensions are more difficult to see in years around solar maximum (i.e., peak of sunspot activity) because of increased airglow. Although there are still questions about its nature, the zodiacal light and its extensions must be primarily caused by dust, and it is the glow of this dust which we, from our vantage point, see roughly marking the trail in space on which our planet orbits the sun. Another, and wonderfully evocative, name for the zodiacal band is the ''light bridge.'' So even in darkest night the sun gently lights the way on which this world of ours will travel, an endless bridge over the gulf of space, connecting the towers of radiance which stand just before the gateways of evening and morning.

The journey of this chapter has been in some respects the journey of the book in miniature—an ever-ascending voyage through a splendor of light in an almost unbelievable variety of colors, forms, and motions. We have seen blue sky, red sun, green flash, purple light, blue-gray of the earth-shadow, silver and blue of noctilucent clouds, gold melting into orange into pink into purple of the volcanically enhanced twilight band in the west, and the many colors of auroras. Light has appeared before us in restful horizontal bands, awesome stationary shadow-beams, a leaping flash or ray at sundown, a double-globe of molten fire at a mirage-met sunset, and fluctuating, racing, rippling, and rotating auroras. We have passed through all these bursts, clouds, globes, and arcs of color and movement on a trip that we can take every clear day and night of our lives, and on which we can be rewarded by both exciting variations and reassuring constancies. This trip unfolds to us our world's pathway through night, from the calm wells of overarching blue to the mystical but ultimately reassuring bridge and towers of comet dust; from the air we breathe and health of home to far reaches of space beyond our atmosphere; from the blue sky to the zodiacal light.

CHAPTER FOUR
Meteors

It was dark, perfectly quiet, and a little chilly as I walked out under the thickly woven branches of the bare trees: the last hour of a night in mid-November. I drew cold air into my lungs, felt its touch full against my face, but my mind was already more than awake, and tingling—with a possibility. I knew there was a chance—however remote—to see a legendary marvel in the heavens that morning. For this was the right morning in 1966 and I knew that at intervals of 33 or 34 years the shower of shooting stars called the Leonids just might possibly overflow the heavens in a display of celestial fireworks like those which blazed in 1833 and 1866 on November mornings when the "stars" fell like snow and the skies were alive with leaping lights. I also knew, however, as I walked under the gloom of the stark trees, that the turn of the century and 1933 had been only disappointments—with little better than the average annual number of Leonids—and most astronomers now believed that the great meteoroid swarm had been shifted away from its thrice-a-century meetings with us by an encounter with the huge planet Jupiter. Was there really any hope on this quiet morning in 1966?

What seemed little hope turned to even less as I came out from under the thick branches: much of the sky was covered with heavy, slow-moving cumulus clouds. It was disappointing: I would be a fairly old man before there was even another ghost of a chance to see this most fabled of all meteor storms. I was ready to give up, but then I noticed that high in the south the mighty constellation of Leo the Lion, source of the meteors, was temporarily clear. I had been gazing at Leo for about ten seconds, pondering the situation, when a meteor—a "shooting star" or "falling star"—hurtled from the direction of Leo's great head, flashed out and was gone in a wondrous instant. As always, the meteor had brought that involuntary catch of the breath and lift of the heart to its observer below. I was thrilled but did not have time to feel encouraged: my thought was interrupted by another meteor. And then another. And not long after by three meteors at the same time. The clouds were large and numerous and morning twilight was already beginning, but it seemed that nothing could quell the meteors that morning. Their ceaseless barrage of fraction-of-a-second streaks and sometimes lingering "smoke" trails gave my wonder no time to rest. Could there be a climax, a still greater marvel than these? Suddenly the sky was rent by the passage of a flaming fireball which fell from on high and seemed to bring the heavens and my breath with it. It disappeared with a burst that lit up an entire area of the sky almost as brightly as a half-moon.

That morning in November 1966 remains in my memory as one of the most exciting displays of natural beauty I have ever witnessed. Seldom have I been so awed—and what I saw was placid and dim compared to what was happening at the same time in the skies of the western United States! What happened that morning was very rare, but the spectacle of dozens of meteors per hour, in all colors and brightnesses, with explosions and lingering luminous trails, is something which any person can enjoy every year, if he knows a little bit about meteors and the annual dates on which they come in showers. Meteors are true visitors from outer space; they can flare to the brightness of the full moon even when they are only an inch in diameter, and are swifter by far than any other objects we see in the vicinity of our planet. They have, on the very rarest of occasions, blasted into Earth with a power and fury which exceeds even the rages of volcanoes and earthquakes. But these falling stars, swift, violent, and profound, also appear as one of the most delicate of phenomena, their strange charm indeed fit to make a wish on or to delight a child. Meteors hold in their shining streaks messages to us from space, but they are also the heavens' fireworks, there for us to enjoy—if we know a little bit about them and when their best shows occur.

In our consideration of these objects, we need familiarity with some special terms of *meteoritics*. *Meteor* is the name given to the astronomical body when it is in our atmosphere and also, more loosely, to the entire light phenomenon (including the trail) which we observe in the sky. Especially bright meteors, those which surpass the brilliance of Venus (our brightest planet), are called *fireballs*. Meteors which explode at the end or in the course of their flight, a single time or many, are known as *bolides*. When a meteor is still in space the proper term for the body is *meteoroid*; if it should manage to survive and reach the surface of the Earth it is then referred to as a *meteorite*. The study of meteorites shows that most of these objects are largely iron or rocky (silicate) or a combination of both. In this chapter, however, we are principally concerned with splendors in the sky, so we will be looking at meteors and *meteor showers*, those concentrations of meteors which appear to radiate out from a single small area of the sky on certain nights of the year. We are primarily concerned with meteors here, but to understand better the behavior and appearance of these objects when they flame in our atmosphere, we do need to begin with a look at them when they are still meteoroids in space.

Meteoroids range in size from the tiniest grains of dust up to the almost mountain-sized masses of the smallest asteroids (there is no distinct boundary in size between the largest meteoroids and the smallest asteroids). Although it is not fully understood, there is definitely a close relationship between meteoroids, the large asteroids (which can be hundreds of miles wide), and comets. Most significantly, we know that many meteoroids are distributed along the orbits of comets, and it may be that most meteoroids are derived from these far larger (and mostly icy) bodies. At least some meteoroids are the debris which escapes from comets when they pass closest to the sun and release their long-frozen gases.

The most spectacular and convincing evidence that meteoroids can arise in this manner was provided by the case of Biela's Comet in the nineteenth century. This object, which returned regularly to the neighborhood of Earth and the sun every 6.6 years, split in two at the return of 1845–1846. Each comet developed its own tail and 6.6 years later the two independent comets appeared again but much farther apart. At the next return (in 1859), conditions were unfavorable for observation, so astronomers waited anxiously for 1866—but in vain: the comet or comets did not show up. Instead there was a notable increase in the number of Andromedid meteors (meteors from the constellation Andromeda) and it seemed very suspicious that these meteoroids traveled in an orbit similar to that of the missing Biela. At the next scheduled return in 1872, when the earth should have passed very near the comets, the comets again did not appear. In their place came a great Andromedid meteor storm (very intense shower) which produced at least 5,000 meteors per hour over parts of Europe. The two halves of Biela's Comet have never been seen again, and though their dark cores may still exist, it is safe to assume that much of the original comet's material flamed in a number of amazing Andromedid storms and showers in the late nineteenth and early twentieth century.

All the meteoroids which are distributed around an orbit are known collectively as a *meteor stream* and the especially thick knots or clusters which may occur at certain points on the orbit are called *meteor swarms*. The most famous example of the latter is the Leonids swarm, which we run into—or sometimes just miss—every 33 or 34 years. One would naturally expect the older streams to have become more evenly distributed (no swarms) along the orbit and to have spread over a wider area, and this seems to be borne out by observations of the showers. Widths of the streams (which are usually measured in millions of miles) determine how long the Earth will take to cross through them and consequently, of course, how many hours, days, or even months the resultant showers will last for observers on the Earth. Since our planet will always pass through the meteor stream at the same place in our orbit, the showers will always take place on the same dates each year. (There are minor changes of the date due to several factors, the most obvious of which is the occurrence of leap years.)

On every night of the year, however, meteors not associated with any known shower may be seen. About six to ten of these *sporadic meteors* can be observed under good conditions (clear sky with little or no interference from moon or artificial lighting). But it may well be that even these sporadics are members of ancient streams which have become very wide and sparse—the final meager remnants of a once-grand comet.

The sporadics can come from any direction, but the shower meteors are distinguished by the fact that they all appear to come from the *radiant*, a point or small area among the stars. The shower meteors are actually arriving in paths parallel to each other, but the illusion that they diverge from the radiant is due to a well-known effect of perspective: in the very same way, we see the parallel rails of a railroad track (or crepuscular rays) appear to diverge from a single point in the distance. Likewise, meteors heading straight at us are seen foreshortened near the radiant as points or flashes of light, and the farther from the radiant a meteor appears, the longer its apparent path. A meteor shower is usually named for the constellation in which its radiant is located or the major star to which the radiant is nearest. Thus the shower meteors which diverge from a radiant in the constellation Perseus are called Perseids (the Greek ending *-id* means "child of").

The dates of the meteor showers are the best times of year to observe shooting stars, but there is also a best time of night to observe them. For viewing meteor showers it is of crucial importance that the radiant be high (the higher the better), and often the position of the moon and the precise time of the shower's peak should be known. But even if all these factors are taken into account, it turns out that, independent of them, meteors are about twice as common after midnight. This is due to the fact that after midnight the Earth's rotation is carrying us forward, in the same direction as the Earth's motion in its orbit. In other words, before midnight the meteoroids that the speeding Earth runs into must catch up to observers, for the world is turning us back away from them. Consequently we see fewer meteors (many fail to catch us), and those we do see appear to be moving slower. But after midnight the Earth's rotational velocity is added to, not subtracted from, the orbital velocity and at 6 A.M. we are at the front edge of our planet, meeting the meteors as head-on as possible. Those morning meteors therefore are faster and, as a result, brighter on the average. Meteoroids in the vicinity of Earth may move at true speeds of about 18.5 miles per second (Earth's speed) to almost 26 miles per second (escape velocity from the solar system), depending on whether their orbit is nearly circular (like Earth's) or nearly parabolic. Meteor showers whose radiants are high in the evening thus have slow meteors, ranging from $26 - 18.5 = 7.5$ miles per second down to $18.5 - 18.5 = 0$ miles per second. Meteor showers whose radiants are high in the morning sky thus have swift meteors which can reach apparent speeds (combination of Earth and meteor) as high as $26 + 18.5 = 44.5$ miles per second. The slowest meteors have their radiant high at 6 P.M., the fastest at 6 A.M. The Leonids are the fastest of the important shower meteors.

Speed is also a major factor in determining the temperature and thus the color of shooting stars. Although most meteors appear whitish (or simply uncolored because too dim to stimulate the cones in our retinas), other hues are quite common. Green is supposed to be the most uncommon major color for a meteor, but I know from personal experience that they are not extremely rare—in fact, several of the most prominent and tumultuous meteors I have ever seen were green. Green is reputed to be the color of many especially bright fireballs. Red, pink, orange, yellow, and blue meteors are all observed. In certain showers particular colors predominate.

As pretty as the countless delicate tints of shooting stars are the trails they leave behind, called *trains*. The light of the trains is not, as it is with the body, primarily the light of friction. Instead, the soft radiance of the trains is mostly the glow of upper-atmosphere gases that have been ionized along the path by the passage of the meteor. The trains reflect radio and radar waves, just like those special layers in the ionosphere which make long-distance radio communication possible. Since the trains are almost all lower than the lowest of the layers (the *E* or *Heaviside layer*), radar can be used to study these wakes of meteors; it is only by this means that daytime showers can be detected. At nighttime, the trains are especially beautiful in certain showers (silvery in the Quadrantids, long-lasting in the Orionids). Persistent trains are associated with bright meteors, and for any given brightness the faster a meteor, the more enduring its train. The trains seldom are visible to the naked eye for more than a full second or so, but in most good showers you are likely to see a few that endure much longer. According to studies, about one meteor out of every thousand has a train which can be viewed to last for more than 10 seconds. On very rare occasions, trains have lingered in sight for minutes, long enough to show the dreamlike wisp change shape and drift off among the stars on the winds of the upper atmosphere.

Meteors very seldom make it farther than the upper atmosphere. Most of them begin shining at an altitude of 80 to 50 miles, and a fraction of a second later (but sometimes longer) they have burned up at anywhere from about 60 to 25 miles above the surface. Fireballs are, of course, caused by bodies slightly larger than the tiny speck which makes an average shooting star, but even most very bright fireballs do not reach the ground. There is a somewhat less remote chance that a brilliant fireball might be audible as it passes—perhaps exploding—miles above. Sound is possible whenever a fireball is less than about 30 miles from the surface, but meteors are so swift that the sound may be heard several full minutes after the meteor is seen. There are possibly reliable reports of "swishing or whistling sounds" heard simultaneously with a meteor's passage. Any lack of sound is easily made up for by the visual splendor of fireballs. Every year there are a number of reports of fireballs much brighter than the full moon and occasionally one is as bright as the sun!

It is by no means certain that even a sun-bright fireball will reach the surface of the Earth. The meteors which do make it are the largest—and the smallest. The tiniest meteoroids are quickly slowed even by the thin air of the upper atmosphere and may float there for a long time, sometimes being concentrated and iced into becoming splendid noctilucent clouds. Eventually these particles are pulled down into the troposphere and come to Earth, often in rain. These *micrometeorites* are a constituent of soil everywhere on Earth. Estimates of the amount of micrometeoritic material reaching the surface vary greatly, but several authorities believe it may be as much as several tons a day. Though apparently a lot, this is still a negligible addition to the bulk of a body as massive as the Earth.

No discussion of meteors is complete without reference to those largest bodies which on the very rarest occasions hurtle down to pound the Earth with catastrophic force. In recent years, searches around our planet have revealed ever more of the eroded remnants of vast craters formed by giant meteor impacts of the distant past. Since three-quarters of the Earth's surface is ocean, we must assume—without knowing for sure—that many more impact craters exist and that our world has been bombarded by immense and unimaginably destructive meteors many times in its history. A few of these bodies would have been better called asteroids, with diameters of a mile and more, and must have utterly devastated large parts of a continent. If they fell in an ocean, perhaps their energy would have been spread more efficiently, causing still greater disruptions (such as mountain-high tsunamis) and possibly evaporating enough water to enshroud the Earth in clouds for years. All such calculations must be rough, but the general effects are all too clear and we are fortunate indeed that the Earth is hit by a very large meteoroid only once in tens of thousands of years. A mile-wide meteoroid probably strikes Earth only once in several hundred million years, but such meteoroids may still have been responsible for mass extinctions such as that of the dinosaurs. The best-known meteor crater in the world may have been formed roughly fifty thousand years ago (though some authorities believe it was formed much more recently). It is the Barringer Crater near Canyon Diablo in Arizona, a hole 4,200 feet wide and 570 feet deep, with a raised rim 130 feet high. It is so perfectly formed that acoustics at its center are uncannily excellent. The object which caused Barringer (or Meteor Crater, as it is often called) is estimated to have been perhaps 200 feet in diameter and possibly to have weighed one million tons. Most of this material was lost in the explosive impact, but a large chunk is believed to lie buried about 1,500 feet below the crater floor.

In the twentieth century there have been no meteor strikes which even approach the magnitude of the Barringer event, but several asteroids have come rather close (just a few times the moon's distance), and in 1908 a region in Siberia was destroyed by what is now strongly believed to have been a piece of comet. (See Chapter Six for an account of this terrible and awesome event.) In recent years the only well-documented case of a meteor striking a person occurred in 1954 when a woman in Sylacauga, Alabama, was slightly injured by a small one which hit her arm and leg. Only a few such cases are known to have ever occurred.

For twentieth-century observers, there have been several mighty fireballs visible over entire states or countries, but perhaps the most impressive display of all was the great procession of meteors observed along a course thousands of miles long on the evening of February 9, 1913. If the sky had not been overcast in New York and New Jersey, several million people might have witnessed the spectacle, but, as it was, observations were mostly from parts of Canada. The procession consisted of many groups of meteors traveling one after the other with a slow and majestic motion across the sky. One of the meteors was as bright as Venus and the total number seemed to have been several hundred. First seen in Saskatchewan, the meteors eventually passed Bermuda and were last sighted from some ships at sea. During this journey, the meteors brightened but only lowered in altitude from about 35 to 30 miles. When over Ontario the groups were separated by distances of 5 to 10 miles and the full procession extended for about 100 miles, taking an incredible three full minutes to pass by observers at any given location.

A convincing explanation for this awesome show was offered by John A. O'Keefe in 1961 (*Sky and Telescope,* January, pp. 4–8). O'Keefe argued that these meteors, which he called the Cyrillids because February 9 is St. Cyril's day, were the fragments of one larger body which had been orbiting the Earth for quite a while as a temporary and very close second moon! As the object's orbit decayed, the large mass heated and broke into the bunches of dozens of smaller meteors, and may have made one last full and very low orbit on fire before burning up completely or reaching the surface in some desolate area.

Although such displays are visible over entire regions of the world (if the weather is clear), you must be lucky to see one. A bright fireball sometimes makes itself known at night by the shadows it casts, which sweep across the landscape in counterpoint to the meteor's motion across the sky (the shadows seemingly race behind trees and buildings to hide from sight of the blazing object). But although great fireballs are the most thrilling of all sights for meteor observers, they are not necessarily the richest or most rewarding. Watching a strong meteor shower at its best is an entertainment that is endlessly varied and surprising. There are always new things to see: meteor behavior of all kinds, bolides and fragmenting meteors, meteors with twisted, crooked paths, and those that emit "sparks." Perhaps you will see even the strangest of all these bodies, the rare *nebulous meteors*, diffuse clouds which may be as big as the moon in apparent size. They have also been called *dark meteors* because their surface brightness is so low that they have been seen silhouetted as they passed in front of the Milky Way.

It is possible just to lie back in a lawn chair or on a ground cloth on a country night and enjoy the show. If you live in the middle of a large city, a trip of maybe only a few miles to darker skies can bring you several levels of enjoyment closer to the starry universe, which is worth the effort on the astronomical holidays of the major meteor showers. In addition to just watching meteors you can also keep a careful tally. For most people, bothering to count the shooting stars is no bother at all; it is instead a way to increase the fun and enjoyment.

Making meteor counts, however, can be more than fun. It is one of the activities by which an amateur astronomer without any equipment can contribute data of considerable scientific value. Your results can be reported to the American Meteor Society (see next page) or to one of the major astronomy magazines (see the Annotated Bibliography for addresses). But in order for your count to be of scientific value, you must at least distinguish between shower meteors (from the radiant) and sporadics (from random directions), and you must keep reasonably accurate track of the time. If you are enthusiastic about keeping a more detailed record of a shower, use a note pad and a flashlight covered with red cellophane (red light will not dazzle your eyes), or, if possible, use a tape recorder to comment on the brightness, color, position, and any notable peculiarities of each meteor. There is great beauty in the solitude of watching the heavens leak meteors in bright breaks of the eternal night's stillness, but if you have a friend or friends who respect the mood it is good to share the experience (and possibly some of the task of tallying or recording details). Keep individual counts and a tally of the total number of different meteors. A more demanding project for two people is to watch the same section of sky and keep careful track of those meteors which both see so that they can afterwards use the *Öpik double-count method*: the total one person sees is multiplied by the total the other sees and then that product is divided by the subset of meteors which were seen by both observers (that is, that were not missed by either observer). The figure derived from the double-count method should be a fairly accurate measure of the most meteors that could have been seen.

From the table and the shower profiles which follow below you can derive almost all the basic information you need to start observing the meteor showers. There are, however, a few important factors which vary from year to year which you need to know about. The phase and rough position of the moon are easy to figure out—even ordinary calendars supply sufficient information. The other important factor (besides weather) is exactly what time the shower will be at its astronomical peak (that is, when the earth will be encountering the most meteoroids, independent of the time of night or height of the radiant at any given location). For many showers this factor is not critical because there is not a sharp peak. But a few showers (notably the Quadrantids) rise to their maximum very rapidly and you will only see peak or near-peak numbers for an hour or two before they dwindle. For information each year on precisely when such showers peak, consult one of the astronomy magazines mentioned in the Annotated Bibliography. Guy Ottewell's *Astronomical Calendar* is another very informative source on meteor showers. (For much general information on meteors, and opportunity to be an involved observer, write to Dr. David D. Meisel, Executive Director, American Meteor Society, State University College, Department of Physics and Astronomy, Geneseo, New York 14454.)

The best single measure of the sharpness of a shower's peak is the duration for which it is above quarter strength—that is,

TABLE 3

Major Meteor Showers

	Dates	Peak	Radiant Position[1]	Radiant Dir.[2]	Radiant High[3]	Duration[4]	Speed (km/sec)	Rate[5]
Quadrantids	Jan. 1–6	Jan. 4	15^h28^m, +50°	north	9 A.M.	14 hrs.	medium 41.5	ZHR 110 40–150
Lyrids	Apr. 18–25	Apr. 21 or 22	18^h4^m, +34°	high south	4:10 A.M.	2.3 days	med.-swift 47	ZHR 12 10–15
Eta Aquarids	Apr. 21–May 12	May 3 or 4	22^h30^m, −2°	southeast	7:35 A.M.	3 days	very swift 67	ZHR 21 10–40
Alpha Capricornids	Jul. 18–Aug. 25	anywhere between Jul. 25 and Aug. 5	(two:) 20^h20^m, −10° 21^h, −17°	south	2 A.M.	——	med. 42	ZHR 6–9 5–30
Delta Aquarids	Jul. 15–Aug. 29	anywhere between Jul. 27 and Aug. 12	(two:) 22^h30^m, 0° 22^h40^m, −16°	south	2 A.M.	——	med. 42	10–35
Perseids	Jul. 25–Aug. 18	Aug. 12	3^h4^m, +58°	north	5:40 A.M.	4.6 days	swift 60	ZHR 68[6] 50–100[6]
Orionids	Oct. 2–Nov. 7	Oct. 20–Oct. 21	(two:) $6^h12.5^m$, +13.5° 6^h25^m, +19.5°	south	4:20 A.M.	2 days	very swift 67	ZHR 35 25; var. 10–70

Taurids	Sep. 15– Dec. 15[7]	Nov. 3– Nov. 13	(two:) 3^h32^m, +14° 4^h16^m, +22°	south	midnight	——	slow 30	ZHR 10–15? 2–10?
Leonids	Nov. 14– Nov. 20	Nov. 17	10^h8^m, +22°	south	6:26 A.M.	4 days	very swift 71	ZHR 5–20[8]
Geminids	Dec. 4– Dec. 16	Dec. 13– Dec. 14	7^h28^m, +32°	high south	2:05 A.M.	2.6 days	med. 33	ZHR 58 50–80
Ursids	Dec. 17– Dec. 24	Dec. 22	14^h28^m, +78°	north	8:25 A.M. (but good all night for northerners)	2.2 days	med. 33	ZHR 9 10–15

[1] Radiant position at peak, in celestial coordinates.
[2] Direction of radiant when at highest viewable for observer at midnorthern latitudes.
[3] Time when radiant is highest (whether observable in dark skies then or not).
[4] Length of time during which the numbers are above one-quarter of the peak rate.
[5] ZHR is Zenithal Hourly Rate; next figure is actual rate observed.
[6] From about 1975 to 1981 the Perseids have produced ZHR and actual rates of 100 and (especially in 1981, it seems) a lot higher; the trend will probably continue until several years after the predicted passage of the parent comet Swift-Tuttle (which is most likely to occur in 1983, though the predictions are highly uncertain).
[7] Over 2 Taurids per hour at best can be expected from October 20–November 30.
[8] Much higher rates possible for some years before and after the passage of the parent comet.
SOURCE: Information in this chart derived largely from Guy Ottewell's *Astronomical Calendar* (which contains much additional material on these and many of the minor showers).

the period during which at least one-quarter the maximum number could be seen. This figure is given in the table. Also given is Zenithal Hourly Rate, which is what it sounds like—the number of meteors per hour which would be visible if the radiant were in the observer's zenith. The column for "radiant position" gives figures in celestial coordinates which astronomers use (see discussion of celestial coordinates on page 235). The showers listed are by no means the only ones known; there are many dozens of feeble and dubious showers, some from radiants that have been determined only by all-night photography, or merely suspected. Among the showers listed, two very strange imbalances are evident, both of which remain unexplained (but may be only coincidental). One is that the majority of these showers have radiants in the Northern Hemisphere of the sky; thus observers in the Northern Hemisphere of the Earth are highly favored. The other anomaly is that most of the major showers occur in the second half of the year.

Perseids. The Perseid shower is extremely dependable and almost always the one which produces the highest hourly rates for a given observer in a typical year. At their best the Perseids may be expected to display about 60 meteors per hour for an observer in dark, clear skies, but the rates are sometimes markedly higher. The trend in recent years has been to particularly high rates (as many as 100 or more Perseids per hour at the peak), which is probably because of the approach of the Perseids' parent comet. Seen only once (in 1862), Comet Swift-Tuttle is likely to return to the vicinity of the Earth no later than 1984 and probably sooner. A close passage of this object could result in magnificent showers or possibly even storms of Perseids.

The Perseids are very swift meteors, many of them bright, fragmenting, or exploding, and often with fine trains. Some can be observed in July, but the shower is above quarter strength from about August 10–14, and the decline from maximum is much quicker than the rise. These meteors are sometimes called St. Laurence's Tears after the famous martyr who is said to have been burned to death on a gridiron (but was probably really beheaded) on August 10, A.D. 258. (It is grimly and beautifully appropriate that meteors are said to be his burning tears.) Laymen notice the Perseids more than any other shower (there are always UFO reports at this time), partly because of their great numbers but also because August nights are a time when many people are outdoors. Weather is usually pleasant for observing the Perseids, though the haze of the northeastern United States is often a problem.

Geminids. This shower is the best rival of the Perseids; it seems quite dependable for a peak of 50 to 60 meteors per hour and may be improving. The radiant is highest (not far from the zenith for United States observers) at about 2 A.M., so that these shooting stars are of medium speed. There is a good percentage of bright meteors and bolides. The average Geminid, like the average Perseid, is about second-magnitude in brightness—not the equal of the brightest stars, but approaching that of the North Star. The orbit of the Geminid stream is very small (the revolution period for an object on it is just 1.6 years), but the parent comet has not been found. It has been suggested that the comet is dead, the largest chunk of its nucleus possibly being the body we call Icarus, the asteroid which approaches closest to the sun. December is a cloudy and cold month in many parts of the United States, but the Geminids are consistently worth braving the chill.

Orionids. This display may produce 30 to 40 shooting stars per hour, but numbers vary greatly from year to year and the shower is above quarter strength for just two days. Many of the Orionids are faint, but about 20 percent of them leave lingering trains (this feature of the shower is very noticeable). The shower is derived from the same meteoroid stream which produces the *Eta Aquarids*: the Orionids are the meteoroids as they pass inward across our orbit, the Eta Aquarids as they pass outward. The Eta Aquarids come at lower rates than the Orionids but are above quarter strength for about three days. It is most accurate to say that the two showers are different substreams of one very large complex. Guy Ottewell points out that Venus also fords through this complex twice in that planet's year. It is sometimes possible for the two planets to be having showers from this same source simultaneously and so have a long, curving (and very sparse) bridge of meteoroids stretching the millions of miles between them. These Orionids and May Aquarids move clockwise in their orbit (as seen from "above"—north of—the solar system). This is unusual because all the planets, all but a few moons, and most comets and meteoroids move counterclockwise about the sun. The Orionids and Eta Aquarids are, however, only partaking in the motion of the comet from which they derive—none other than Halley's Comet itself. Therefore, when you watch the Orionids or Eta Aquarids you are actually seeing tiny fragments of that famous comet. Both of these showers might be enhanced for years before and after by the passage of Halley in 1985–1986.

Delta Aquarids. This shower can be confused with several nearby meteor showers which occur at about the same time and which together with the Delta Aquarids aid the Perseids in making the period from late July to early August the most plentiful for meteors—a midsummer meteor maximum. The Delta Aquarids are more widely and evenly spread than most major showers, so that the peak date can fall anywhere within a period of about two weeks during which numbers are high. Therefore a bright moon can never utterly ruin observations of this shower, which features comparatively slow meteors (a nice contrast to the concurrent swift Perseids). The total number of these often yellow meteors exceeds that of any other shower, but since the peak is not sharp, the best hourly rates average no more than about 20 per hour at best. The nearby *Capricornids* are also evenly spread, yellow, and may peak at the same time, when they can add as many as 10 or 15 shooting stars an hour to the total coming from this region.

Quadrantids. These meteors can be even more plentiful per hour than the Perseids, but the peak of the shower is extremely narrow; in many years the peak occurs while the radiant is low or in daylight, depending on the observer's location. The Quadrantids are above quarter strength for only 14 hours, so in certain unusual years you may be fortunate enough to see 100 or more an hour, but in most years you will miss the peak period and see few. These frequently blue meteors leave silvery trains. They come from an area of the constellation Boötes that was formerly considered a separate constellation called Quadrans Muralis, the Wall Quadrant, named for a mariner's measuring device.

Fig. 20. *The Leonid meteor storm of November 17, 1966, showing two point meteors near the radiant, and many other Leonids. Seventy Leonids were recorded in the $3\frac{1}{2}$ minutes of this exposure, whereas during most very good showers a photographer is lucky to catch even one meteor bright enough for its brief life to make enough impact on the film. Photographed by Dennis Milon, of the team of amateur astronomers who saw over 1,000 meteors per* minute *during a 40-minute period centered roughly on the time of this photo.*

Taurids. This shower lasts for about three months and comes to a gradual maximum around early November, when no more than about 10 meteors per hour might be seen. These very slow meteors, derived from Comet Encke (see Chapter Six), include many fireballs. A slow, bright shooting star seen in late autumn usually turns out to be from Taurus.

Leonids. The swiftest of shower meteors, the Leonids are not numerous in most years, but the stream contains one swarm that has produced the greatest meteor storms ever recorded. Every 33.17 years the comet (Tempel-Tuttle) with which the

Fig. 21. *Leonid meteor storm of November 17, 1966, showing star trails (caused by the Earth's rotation during the several minutes' exposure) being crossed by the longer trails of Leonids far from the radiant. Sirius, the brightest star, appears to be the brightest object in this photo, and the bright Orion the Hunter is also visible, but some of these meteors must have been in reality far brighter to show up as they do, though lighting the film for a mere second or less. Photographed by Dennis Milon on Kitt Peak, Arizona.*

Leonids are associated returns and is followed by an intense swarm which the Earth sometimes enters that year or the next. The displays in 1799, 1833, and 1866 filled the skies to overflowing. The greatest of these three may have been the storm of 1833: in Boston, the flashes were so bright and frequent that they awakened people inside their homes and one fireball was nearly as large as the moon; it was the night the "stars fell on Alabama," when the maximum count was of 35,000 meteors per hour! After the great 1866 storm the next two returns of the comet produced no spectacular Leonid activity and expectations were not high for the 1966 shower. You will recall that I began this chapter with reference to my own experience that year and said that my clouded, increasingly twilit view, unforgettably good as it was to me, was yet only a minor glimpse of the awesome glories which filled the skies further west.

The 1966 storm rapidly peaked after daylight came to the eastern United States, when the radiant was highest for the Southwest. It turned out to be unquestionably the greatest meteor storm we know of in history. In about two hours the rate at Kitt Peak in Arizona rose from a mere 33 per hour to 30 *a minute* (1,800 per hour) . . . and then the full fury of the storm broke. So plentifully did the Leonids pour that many observers, though they knew they were not in danger, could not help involuntarily shielding themselves from the apparent onslaught. Such a torrent of meteors burst forth that for some observers the entire Earth seemed to be moving: they were overwhelmed by the impression that the Earth was a ship hurtling toward the radiant through still meteors (how amazing considering how fast the Leonids are!). The meteors fanned out in a huge billowing umbrella of light from the radiant. In some directions observers said the meteors seemed to rush in such a constant wave that it looked as if a waterfall of shooting stars was flowing down the sky. A one-second sweep of the head revealed 10, 20, and even 40 new meteors every second; shadows and multiple shadows were cast and raced again and again; twisted trains of glowing meteor smoke were strewn across the sky. At Kitt Peak meteors fell at an estimated rate of 150,000 per hour for about 20 minutes, and there were spates as high as 140 meteors per second—504,000 per hour!

The next Leonid meteor storm may take place on November 18, 1999, but other storms which have produced thousands of meteors an hour from elsewhere in the sky have happened without warning on a number of occasions. The opportunity to witness the heavens streaming with the streaks, sparks, smokes, flares, and rushing fires of a torrential meteor storm is always a possibility when you sit out in the still night to watch a shower of shooting stars.

II. THE SOLAR SYSTEM

CHAPTER FIVE
Eclipses

Though it was the middle of a cloudless day, the temperature had plunged about 15° in the last hour. Across the small town and the inlet by which I stood, the light had taken on a strange cast and had lost most of its power. Glitters in water and car windshields once bright were now feeble, and the reeds, grasses, and bay had deepened into more somber hues. Along this stretch of beach with me were many people (perhaps a few hundred) who had been very noisy until a few minutes before. Now they were hushed, silent, as was the entire land for miles around. All things seemed poised, waiting for some stroke in wonderful dread. Suddenly there was a shivering sense of change on Earth and in the heavens, like the first whisper of a great wind's roar heard coming from afar at the break of a storm. And then, the final fall: perhaps ten or fifteen seconds in which darkness, dozens of times vaster than any storm's, hundreds of times swifter, rushed over us. In that darkness I stood transfixed, my horizon ringed with the deepest orange-red glow, with Venus beaming in a midnight-blue sky, beside the sight which some people have called "the eye of God": the seemingly jet-black disk of the moon surrounded by pearly shimmering, the gentlest light I have ever seen—the corona of the sun. For just under two minutes I stood there, more than what we call a person—or perhaps I was the pure core of wonder which lies buried deep in the mind of every person. I was witnessing a total eclipse of the sun.

There are several varieties of both solar and lunar eclipses, and though none is so grand and literally stunning as a total eclipse of the sun, all are at least a little eerie and more than a little interesting. A total eclipse of the moon is an artist's dream: its unpredictable and various tints develop slowly enough to luxuriate in, to recall in detail, and perhaps to sketch even as the eclipse unfolds. Eclipses are such remarkable sights because they are drastic changes in the appearance of the heavens' two most splendid bodies and the entire environment of sky and earth affected by those bodies. We marvel at a shooting star, a comet, a conjunction, a rainbow—sights which we cannot see all the time—but the sun and the moon, whose beauty and power are greatest of all, are visible so often we usually take them for granted. Familiarity may not always breed contempt, but in this case it dulls the eye of wonder. Eclipses show us the familiar faces of the sun and moon in different guises and help to restore the kind of wonder which we should have (and did have when very young).

The sun, Earth, and moon are in line at every new and every full moon, but they are not always in the near-perfect line required to cause an eclipse. This is true because the moon's orbit is slightly inclined to the plane which contains sun and Earth. In other words, at full or new moon, old Luna is sometimes a little above or a little below the position that would make a perfect line. An eclipse can only take place, if the moon is new or full, when it is also at (or near) one of its "nodes," these being the two places where the moon crosses the sun-Earth plane.

For predictors of eclipses there are many additional factors to take into account, but the overall result is for each year to have no more than seven eclipses (of which four or five may be solar and three or two lunar) and no fewer than two (both of which are then solar). Despite these figures, an observer at any single location on Earth will see lunar eclipses far more often than solar, because eclipses of the moon are visible from an entire half of the planet, whereas solar eclipses are much more localized (because they are caused by the moon's shadow falling on Earth, and that shadow always covers far less than a full hemisphere). Very localized indeed are total eclipses of the sun, for the central shadow of the moon is never more than a few hundred miles wide as it sweeps across the turning Earth. Consequently, a total solar eclipse occurs on the average just once in about 360 years for any given spot on earth, so you will probably have to travel a little if you want to see this grandest of all celestial events.

Before we explore the wonders of solar eclipses, let us first consider the less spectacular (but still uniquely beautiful) eclipses, those of the moon. Lunar eclipses are caused by the Earth's shadow, which (like our own shadows) is composed of two parts. The inner, darker region of the shadow is the *umbra* ("shade," as in *umbrella*); the outer, less dark area is the *penumbra* ("almost shade"). The long tapering cone of the Earth's umbra extends almost a million miles from earth, but the moon's average distance from us is only 235,000 miles, so at that point the umbra is still over 2½ times as wide as the moon, while the penumbra is over 4½ times the lunar diameter here on Earth. We can imagine the cross-section of these cones as circular areas in the sky, the penumbra concentric about the umbra, with the full moon usually passing either just above or below them. On some occasions, however, the moon will pass into them and we will experience an eclipse of the moon.

Depending on how deeply the moon penetrates, there can be three types of lunar eclipse: total, partial, and penumbral. The least interesting is the last of these. It is quite possible for the moon to pass through the penumbra without ever reaching the umbra, in which case the lunar surface is only slightly dimmed and the event is called a *penumbral eclipse*. When the passage is almost tangential, so that only a small portion of the moon brushes through the outer fringes of the penumbra, then the dimming of that portion will be too slight to be detected by the eye. Only if about 50 percent of the lunar diameter is covered will the edge of the moon farthest into the penumbra be stained deeply enough to make a distinct contrast with the still unshadowed area of the lunar face.

Whereas the penumbra is hardly noticeable, the umbra is much darker and becomes visible to the naked eye soon after it first touches the moon, beginning the *partial lunar eclipse*. As you watch this dark shadow creeping, minute by minute, across the moon's bright face and follow its progress in relation to the major lunar markings, take careful notice of its shape. You can see for yourself that the edge of the shadow is curved and know directly that the planet you stand upon really is spherical.

As the partial eclipse advances, reversing in minutes with

Fig. 22. *A partially eclipsed moon gives way to dawn and the Grand Tetons in Wyoming.*

oddly shaped phases two weeks of waxing, one of the most beautiful effects is the darkening of the sky and apparent brightening of all the stars. As more and more of the moon's surface is dimmed, it is as if a greater night were falling upon night, as if some mysterious power were turning up the fires of the stars, kindling them to finer radiance. Hundreds of previously too-faint stars come into view.

How dark the sky really does get and how the stars appear depend on the darkness of the individual eclipse. Some people who have never seen a *total lunar eclipse* believe that when one occurs the moon, like the face of the sun in a solar eclipse, is simply blotted out. In reality, that is very seldom true. Although the moon's brightness is always greatly diminished in total eclipse, it is usually still greater than that of the bright stars. What is amazing and lovely is that this dimmed moon shines with a *red* light.

What is this strange red light which glows from the lunar face in total eclipse? If the Earth could be observed from the surface of the moon during an eclipse, the dark bulk of planet Earth would easily be large enough to cover the globe of the sun; so you might suppose that only starlight would be falling on the lunar surface. But that is not so, and the reason is the Earth's atmosphere.

Our observer on the moon would see the large dark shape of Earth (whose phase would be "new Earth") move in front of the blinding ball of the sun, and then a glorious ring of light would appear around the Earth. This would be light passing through the Earth's atmosphere and refracted into the Earth's umbra. The light is red for the same reason it is so when we see it in a sunset, except in this case the two passages of light (in, then out of our atmosphere) make it even ruddier and more dimmed. When it leaves our atmosphere it is destined for the now otherwise dark lunar surface, but for us to see this light it must of course make the journey back to plunge again through our atmosphere, this time reaching our eyes. If this elaborate journey were always exactly the same and always produced the same hue of red, it would be marvelous enough. But the Earth's atmosphere, from bottom to top, is a remarkably changeable and complex medium, and the result for eclipse observers is an astonishing variety of tints and degrees of brightness from one eclipse to the next.

The brightness and color of total lunar eclipses ranges from brilliant orange (which some casual observers might hardly notice), through ever deeper reds (some of which surely terrified the superstitious of past ages with the fear that the moon was bloodied), and finally down to rare dusty gray or black eclipses when the moon has actually disappeared. The following scale, devised by A. Danjon and now in popular use, can be used to rate the brightness and color of total eclipses of the moon:

L = 0. Very dark eclipse. Moon practically invisible at mid-eclipse.
L = 1. Dark gray or brownish eclipse. Surface features difficult to make out.
L = 2. Dark red or ruddy eclipse, frequently with a large dark patch at the center of the umbra, surrounded by a slightly brighter outer zone.
L = 3. Brick-red eclipse. Umbra frequently bordered with a gray or bright yellowish zone.
L = 4. Coppery red or orange eclipse, very bright. Luminous outer zone of a bluish tint.

Since the umbra is darkest at its center and is far larger than the moon, it follows that one reason for a brighter eclipse could simply be the moon passing (still in its entirety) through the outer, lighter part of the umbra. Indeed, midtotality usually is quite a bit darker than the beginning and ending phases, when not only a "luminous outer zone of a bluish tint" but many other colors may be seen on the moon (these are often, but not always best observed with binoculars or telescope). Minnaert (p. 295) speaks of the outer zone of the umbra sometimes consisting of "rings coloured successively bright sea-green, pale golden, copper, peach-blossom pink (from the inner part outwards)." What a treasure for an artist's canvas, and all of it ever changing, though changing gradually enough for quick sketches to capture it and for any enthusiastic observer thoroughly to enjoy it. But it is not primarily how centrally the moon passes through the umbra which determines the colors and intensity of illumination on the eclipsed moon, it is the state of the Earth's atmosphere in the ring dividing day and night around our sphere.

The ring around the Earth is the *sunrise-sunset line* or *terminator*. The areas of the world that are experiencing sunrise and sunset during the period of total eclipse are the places whose atmospheric conditions will influence the illumination and color of the umbra on the moon. Indeed, since the sun's light is inevitably bent around the Earth by our refracting atmosphere, it is quite possible to say, as Minnaert does, that the Earth's umbra cannot in all justice be called a shadow, since it is really a beam of delicately tinted light cast out into space by our ring of refracting atmosphere. Minnaert (p. 295) goes so far as to make the deliberately shocking statement that "it is quite impossible for the shadow of the Earth's globe to cause an eclipse. . . ." But of course we could never see the delicate colors refracted into the umbra amidst the general flood of sunlight in space if the bulk of the solid Earth were not shading out the fierce glare from the long tapering cone of the umbra.

What then are the conditions of the atmosphere along Earth's terminator which will lead to a particularly dark lunar eclipse? An unusual amount of cloudiness throughout this zone would prevent passage of the light that goes through the lowest level of the atmosphere. But to produce a very dark eclipse it is also necessary for the higher regions of the atmosphere, far above clouds of water or ice, to be less transparent than usual. That condition is met when very violent volcanic eruptions discharge vast quantities of ash to float for months in the calm of the stratosphere. I remarked earlier that the moon sometimes does completely disappear during a total lunar eclipse, but you may have noted that the darkest rating on the Danjon scale only has the moon "practically invisible at mideclipse." The scale was invented, however, before the widely observed eclipse of December 30, 1963, which vindicated the claims of observers of eleven previous very dark eclipses since 1601. On that night, as I saw for myself, the moon really did disappear to the naked eye (and even to some telescopes) during parts of totality. The totally eclipsed moon, which is usually brighter than even the bright stars, was rated as fainter in magnitude than an average star. And this brightness was not concentrated into a single point but spread out dully over the large area of the lunar disk. No wonder this sooty smudge of a moon was sometimes lost to view, haunting a brightly starred sky like a phantom.

Not since 1913 had so dark an eclipse occurred, but, as in 1913, there had been recent major volcanic eruptions (Mount Agung in Bali erupted earlier in 1963) which had caused spectacular sunsets across large areas of the world. Weather satellite data showed that there had also been an unusual amount of cloudiness along the terminator. As this book goes to press, the moon in the July 6, 1982 total eclipse was partly blotted out by the effects of the El Chichon volcano ash-cloud, and the December 30, 1982 total lunar may turn out to be as dark as its predecessor of exactly 19 years earlier. It is fascinating that today, as in 1963, it is still usually more difficult to predict the specific degree of darkness and the regions of color on the eclipsed moon than it is to gain information about our atmosphere from studying the eclipse. It is quite possible to determine which precise areas of the umbra will be affected by very specific regions over the Earth. Thus it was possible for one expert in 1963 to relate a red area near the moon's north rim with clear skies west of Sumatra and to infer confidently from a bright blue fringe on the moon at one point the existence of volcanic dust not far west of South America at a certain latitude!

In general, the light passing through the lowest layers of atmosphere is refracted most and thus is found at the center of the umbra. Those beautiful fringe colors mentioned by Minnaert are likely to have been caused by the uppermost atmosphere; spectroscopic studies of this light coming back from the eclipsed moon have taught us much we would otherwise not know about those mysterious and most ethereal regions of our atmosphere. One people of central Asia believe that the moon is a mirror reflecting in its face all that happens here on our world. Miraculous as it may seem, there really is a grain of truth in that assertion: who would have guessed that we could learn about the inaccessible regions 50 and more miles up, about the clouds of tiny volcanic dust roving our atmosphere, about weather in specific areas of the Earth many thousands of miles away, about the very shape of our vast planet—all from looking at the moon in eclipse?

There are seven precise moments of special importance whose times of occurrence are predicted for each total eclipse of the moon (fewer of these apply to the partial and penumbral eclipses): moon first touches the penumbra; moon first touches the umbra; last of moon enters umbra, totality begins; *mid-eclipse:* middle of totality; moon begins to leave umbra; last of moon leaves umbra; last of moon leaves penumbra. In an eclipse where the moon passes centrally through the umbra, the entire time from first to last umbral contact can be as long as about 4 hours. The July 6, 1982 lunar eclipse featured the umbra touching the moon for about 3 hours and 56 minutes and a totality of just over one hour and 46 minutes—the longest lunar totality in more than a century, and the longest visible from the United States since before the country's founding!

It is not surprising that lunar eclipses have provoked great fear in cultures which did not understand the cause of the phenomenon. One such culture was the Indians of Jamaica, whom Columbus awed with his accurate prediction of a total lunar eclipse. In 1504, the great explorer was stranded and in desperate need of help from the unfriendly natives. Columbus had noticed in his almanac that a total eclipse of the moon was going to occur. He used his foreknowledge of the eclipse to convince the Indians that he was in contact with the great one god and therefore must be honored and obeyed by them. The ploy worked. (Columbus also used the event as a means of determining the longitude of Jamaica.) Lovely, eerie lunar eclipses have played a role in many other historic happenings—from the greatest disaster in the history of Athens (the city's defeat by the Syracusans, which led to most of 7,000 Athenians being enslaved to death and was occasioned by their superstitions about the total lunar eclipse of August 27, 413 B.C.) to Ben Franklin's discovery of the counterclockwise airflow of hurricanes—and they have left their mark in legends, the collective imaginations of many cultures.

But even the loveliness of the total lunar eclipse is only a dim cry of that shattering beauty which few people have seen, "the coming of the sudden shadow": the total eclipse of the sun.

Eclipses of the sun are of course caused by the moon passing between the Earth and sun so that we are the ones who have a shadow (the moon's) cast upon us. The moon's diameter, however, is only about one-quarter that of our planet; consequently, even at less than our average distance from the moon the cone of shadow of a solar eclipse tapers down to only dozens of miles wide. If the eclipse occurs where the sun is overhead, then the cone of shadow is cut by the plane of the Earth's surface perpendicularly (except that the Earth's surface is curved) and the shadow (as seen from far above the Earth) is almost circular. In most cases, of course, the eclipsed sun is not directly overhead, so the shadow cone is slanted and the cross-section cut by the Earth's surface is an ellipse. Satellite photographs taken during eclipses have revealed this dark ellipse of umbra-covered area and shown it surrounded by a far larger region of lesser shade (thousands of miles wide), which is obviously the penumbra of the moon's shadow. Whereas observers in the ellipse are seeing the moon's dark form fit perfectly in front of the sun's entire blazing disk, people standing in the penumbra can only see (with protective devices or other safety methods) part of the sun's disk covered. In other words, when we stand in the moon's penumbra, we experience a *partial eclipse of the sun*.

The view from space at any given moment on the day of a total solar eclipse would show us the black umbral area and surrounding penumbra on Earth, but the situation is by no means static. The moon moves onward in its orbit and the Earth itself is rotating. The result of this complex combination is a *path of totality*, which is the course of the umbra across thousands of miles of land and sea, beginning at sunrise in one part of the world and (usually) running basically east to a place where the eclipse occurs at sunset. Though the path of totality might seem long, the vast majority of the Earth's surface area is either open sea or nearly inaccessible wilderness such as Siberia. Fortunately, in recent years air travel and organized eclipse tours for the public have made total solar eclipses available, though still not inexpensive, to many more people. And, of course, for several thousand miles on either side of the long, narrow path, there is the opportunity to observe a partial solar eclipse.

Even rather small partial eclipses of the sun are dramatic to the observer who understands what is happening, especially if he can view the eclipse by the proper safe means. Many people are confused about whether it is safe to look at solar eclipses and most seem to think that it is not. The simple rule to remember is that the totally eclipsed sun is safe to look at, but the partially eclipsed sun is dangerous: even a small bit of uneclipsed solar surface could seriously damage your eyes in a few seconds of staring. How then can a person observe a partial eclipse safely? Some people try to use dark photographic negatives or smoked glass but very few such materials really block out all the harmful rays. Not even all kinds of welder's glasses are safe for looking at the sun, and mylar must be used in the proper thickness. There are filters which are specially designed for observing the sun, but even with these there is possible danger for the inexperienced. The safest and easiest way to observe partial eclipses (or the sun at any time) is unquestionably by *projection*. This technique is simply viewing a projected image of the sun, which may be accomplished satisfactorily with equipment as technically unsophisticated as a piece of cardboard or a tree! If you have no other safe optical aid, take a piece of cardboard or any opaque material through which you can punch a pinhole. The size of your pinhole and the distance from the cardboard to the surface onto which you project will determine the size of the solar image and its sharpness. All you need do is hold your cardboard perpendicular to the rays from the sun and look at the little circle of light, which actually is a sharp image of the sun. The chinks between leaves of a shade tree perform the same projection in nature, and if you look closely in the shade of almost any tall leafy tree, you will see that the dappling of the ground with sunlight really is in the form of sun-images of all sizes and sharpnesses—but most, of course, are elliptical rather than round, unless the sun is very high.

The sun-images formed by trees and pinholes are only rarely capable of showing detail such as sunspots. (Imagine looking at a sunspot—a surface feature on a body 93 million miles away—on the ground beneath the shade tree in your back yard!) But these images are sufficiently large to show you the precise development of a partial eclipse, for as the moon begins to move in front of the sun, a bite is taken out of the edge of the sun-images and grows as the eclipse progresses. With binoculars or telescope, projection is accomplished by simply focusing the instrument to form the image at the distance of the surface to be used as a screen. Using optical aid provides a much larger, more detailed image; with a telescope you can typically see a number of sunspots and sunspot-groups and, during a partial solar eclipse, watch the rounded edge of the moon creep across to block off the view of one spot after another.

As a partial eclipse increases, it is not only the sun and moon which should be observed but the entire environment of earth and sky, something which can only be done properly with the naked eye, and with all the senses. Most authorities say that the reduction of illumination becomes noticeable when the moon has moved across about half of the sun's diameter, but to the careful observer who knows what to look for, the changes are visible much earlier. For instance, anyone who has studied the blue sky will notice the darkening of the sky even very early in an eclipse. Indeed, photographs taken at different early stages of the partial eclipse will show vividly the changes in the sky and landscape long before the sun's width is half-covered.

As the moon covers still more of the sun, the rate of visible changes accelerates. When about 70 percent of the sun's diameter is hidden by the moon, the lighting becomes peculiar for a new reason: the sunlight reaching us is coming from an edge of the sun and that light is both less intense and *redder* than the light from the center of the sun's disk.

You are most likely to experience all the exciting phenomena which occur at a 90 percent partial phase as a prelude to the total eclipse which you have traveled to witness, so it is most appropriate to discuss those sights in the context of observing a total eclipse. But there is one other major kind of solar eclipse which may follow the partial phases and which can be spectacular, even if much less so than the terrible splendor of totality. An *annular eclipse* takes place when the moon passes centrally in front of the sun but is too far from the Earth at the time to cover completely the sun's blinding disk. The orbit of the moon is not perfectly circular and only if the moon is at less than its average distance from us will it appear large enough to cause a total eclipse. Otherwise, the eclipse can only be, at best, annular: a ring (Latin *annulus*) of the sun's bright surface is still visible around the dark moon. Although that ring is very narrow, it is still too bright to look at safely and prevents the great darkness of totality from falling (the umbra of the distant moon has tapered to a point just before reaching the Earth). Rarely, there is an *annular-total eclipse*, in which the umbra reaches the surface of the Earth for just a part of the entire central path of the eclipse.

The possibility of both annular and total, and even an intermediate type, indicates how remarkably close the fit between sun and moon is. It is an unusual coincidence which probably has not always existed and probably will not always continue to exist in the Earth-moon system (various forces lead to changes in the distance between Earth and moon). On other planets and moons of our solar system, there are other such coincidences which lead to spectacular sights, but the one we are blessed with is especially fortunate: our moon is often big enough to block the sun's disk, but not so big as to block the beautiful layers of the solar atmosphere just above the surface of the visible disk. We consequently get a chance to look at these more dimly glowing regions which are normally altogether overwhelmed by the blinding radiance of the sun's surface, its *photosphere* (*photo* means "light"). The largest of these outer regions is the lovely *corona* ("crown"), which must not be confused with the cloud-caused coronas we looked at in Chapter One. There is an inner and outer solar corona and if the moon appeared much larger from Earth, the inner corona would be covered up, too, and total eclipses would be somewhat less beautiful, though they would last longer.

Actually, it should often be possible to get a dim glimpse of the corona during an annular eclipse, but one would have to be extremely careful (observers have sometimes spotted the corona even a few minutes before totality). I have never read about what the sun-images and shadows look like during annularity, and I have not yet seen one of these uncommon eclipses myself, but surely image and shadow must become ring-shaped—a truly bizarre sight! Many of the striking phenomena visible just before or after totality can be seen at an annular eclipse. The annular stage can last as long as about 12 minutes and 24 seconds—brief indeed, yet substantially longer than totality can ever be.

No person who has ever witnessed totality would trade it for any duration of annular eclipse. A total eclipse of the sun may be the most awesome sight in all of nature; certainly it is the most wondrous that occurs rarely but regularly, making prediction possible. High in the sky, down on the horizons, even on the ground—everywhere, wonders appear, transform, and race on in a matter of seconds in that enchanted time out of time. To convey some sense of the wildness, rapidity, and richness of these sights and sensations, it is probably best to tell of my personal experience of one of these singular events. I witnessed an eclipse in the tiny village of Lundar, in the middle of Manitoba, Canada, on the morning of February 26, 1979—a day which to me is unforgettable, yet no more so than

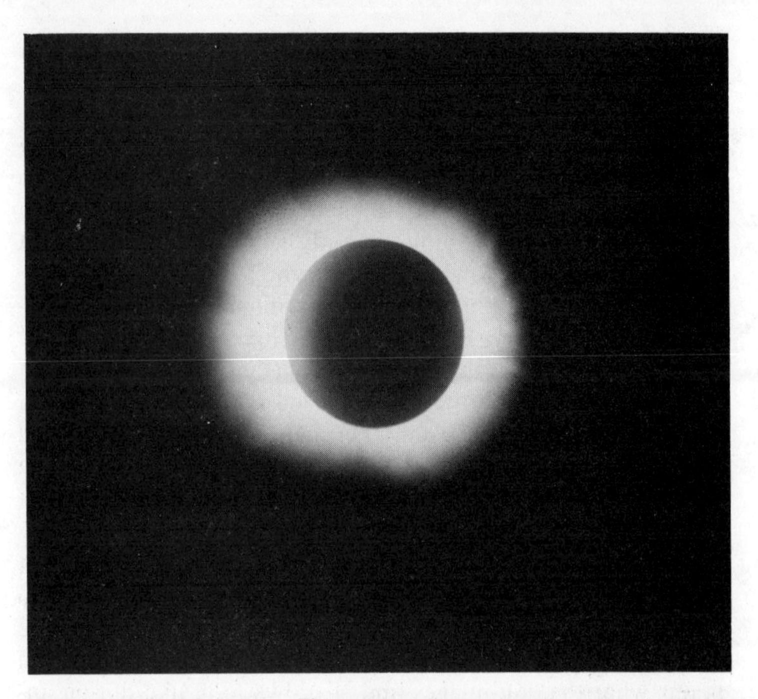

Fig. 23. *Total solar eclipse. The corona is larger than it appears here, only the brighter middle and inner coronas showing in this moderate-exposure photograph taken in Saskatchewan. (See inside back cover.)*

that of the other total eclipse described at the opening of this chapter (and with which I will supplement this account).

As the minutes dwindle down toward the beginning of totality, the eerie dimming of light has become noticeable to everyone, and also to birds, insects, and other animals. The coming of the moon's shadow may bring with it the *eclipse wind*, an often strong gust caused by a meteorological factor that will always be felt if skies are reasonably clear—namely, the drop in temperature. I experienced a large drop of temperature at my earlier eclipse, but at Lundar, where light cirrus clouds feathered most of the southern sky, the thermometer fell less—fortunately, because even without the eclipse the temperature was well below freezing—and yet this was one of the warmest days of the winter in Manitoba! Despite that cold, I was just one of tens of thousands of people who had traveled to this province so decidedly out of season—for a few minutes of glory in the skies.

The light cirri about the sun made it possible to glimpse the remaining thin slice of solar crescent vaguely in a swift glance, and now, while all waited for the solar corona, a murky-red cloud-corona dimly circled the sun. Little did I suspect that this cloud-corona was a herald of a totally unexpected sight that was soon to appear—a sight more miraculous than any single one I have ever witnessed in the heavens.

There were snow and ice on the streets and fields, a good viewing screen for a phenomenon which I did not see, however, at this eclipse: *shadow bands*. These bands are strips of shadow, usually about an inch to a foot wide and somewhat farther apart, which race across the landscape at fairly low speeds. The probable cause is perhaps as eerie as the sight itself: we are watching the effect of turbulence in the atmosphere, which makes the narrow solar crescent twinkle and thus produce alternations in the illumination. The shadow-band phenomenon is sometimes visible on a white surface in a dark room into which is admitted the light of a very bright star, or it can be produced more readily by using a light-colored surface (a shirt will do) to catch the bands caused by distant car headlights shining through the unsteady air over a road that has been heated in the sun all day. These experiments are, of course, only feeble relatives of the eclipse shadow bands, whose screen is not paper or a shirt but huge areas of landscape and all the objects in it. It certainly is strange to think of the *sun* twinkling; obviously the solar crescent must be very thin, and total eclipse near, for shadow bands to be seen (though they have occasionally been spotted quite a few minutes before totality). They are not always visible at a given site for a given eclipse, but when they do appear they may be very prominent indeed. In an article in *Sky and Telescope* magazine, there is a delightful report of shadow bands apparently so noticeable that they were followed with interest by a dog!

The most awesome of all the phenomena before totality is the approach of the shadow itself. It may be very difficult to observe if the sky is clear, but very light clouds can render the oncoming umbra spectacularly visible. It is an imposing sight. It appears much like a thunderstorm in color and darkness, but this "storm" is not a few miles or dozen miles in diameter or traveling at 10 to 60 miles an hour. In the case of the eclipse at Lundar, the shadow I saw approaching was about 192 miles long north to south and about 85 miles east to west. The first of these figures is abnormally large for the umbra, but the speed at which this immense darkness was moving was fairly typical—about 1,800 miles per hour!

The shadow is so dark that with any help at all from clouds or dust to make it visible, it can be sighted when still very far away. The earliest sighting before totality which I know of was in the great African eclipse of June 30, 1973, where one observer is reported to have spotted the shadow 8½ minutes before second contact. (In a solar eclipse, second contact is the moment when totality begins; it ends at third contact. The moon is first and last touching the sun at first and fourth contact.) Another impressive early sighting of a faster and therefore quite distant shadow was made at the 1979 eclipse by a friend of mine from a site in Saskatchewan. He observed the shadow 2 minutes and 35 seconds before second contact and 3 minutes 58.5 seconds after third contact. In both cases he could have observed the shadow when it was even farther away if he had not been busy with other observational activities, but when he last saw the moon's umbra its rear edge was on the ground about 120 miles to the northeast, in which direction the shadow's forward edge was simultaneously causing totality for me far away in Lundar.

It seemed to me that the shadow was visible for roughly as long at Lundar as it was for my friend in Clarkleigh, Saskatchewan. The distant visibility of the shadow is of special interest to observers who might not be able to get into the path of totality but have to settle for seeing a very large partial eclipse from many miles outside the path. If you ever find yourself in this situation keep your eyes open for the shadow passing along the horizon. Whether you are inside or outside the shadow, by the way, a mountain site can provide some magnificent views of the shadow.

Another point about visibility of the shadow in seemingly adverse conditions is very important: that you do not presume that the coming of totality will not be detectable if skies are overcast. The umbra is very dark and, contrary to expectations, overcast skies can sometimes make the *arrival* of the shadow even more dramatic, because low clouds will block the view of the approach until the umbra is almost upon you. Some observers have said that the shadow rushing over then is like a shutter closing across the sky!

In addition to the awesomeness of the eclipse shadow's speed, darkness, and vastness, there is the wonder of what it is: the end of a dark cone almost a quarter-million miles long, the shadow of that very solid and immense world which we have always seen as a friendly and fairly small disk of light in the distant heavens.

As I watched the wall of shadow climb in the southwestern sky at Lundar, now accelerating as it grew near, my eyes were drawn back toward the sun by a nearby person's shout of "Baily's Beads!" These "beads" (which is what they do look like) are remaining sections of the sun's surface visible after the last, thinnest sliver of solar crescent is broken into pieces by the unevenness of the moon's edge: we are watching sunlight pouring through the moon's lowlands seen in profile. The beads are not too bright to look at and were actually noted long before Francis Baily wrote of them in 1836 (for instance, by Halley at the eclipse of 1715). Here at Lundar I saw three beads in those last seconds, as the final, most precipitous fall to darkness began, the corona becoming dimly visible around the moon's disk. The sky-spanning forward edge of the shadow was now just a breathless heartbeat away from the sun.

Totality. Full darkness. There is a wonderful simultaneity of the shadow's edge reaching the sun and the solar corona suddenly springing into full blossom. How dark does it become? It varies greatly from one eclipse to another (and the topography and reflectivity of the site region makes a difference), but the illumination is roughly comparable to the middle of twilight, or a little brighter than a night with a full moon. This may not seem extremely dark, but go out on the next full-moon night and imagine the lighting of a very cloudy day falling to this level in a matter of seconds. For both psychological and physiological reasons, the precipitousness of the eclipse darkness makes everything seem much darker than figures suggest or an observer would expect. At Lundar I am sure the totality was somewhat lighter than the one I had observed at my earlier eclipse, but people who were seeing a total eclipse for the first time were astonished by how dark it got.

An impressive and wonderful measure of the eclipse darkness is the appearance of bright planets and stars during totality. Venus is ordinarily so bright it can often be observed with the naked eye in the bright sunlight of a clear day, but only if you know exactly where to look. During a solar eclipse, however, Venus becomes prominent a few minutes before second contact and beautifully bright during totality. You are only likely to identify at most a few bright planets or stars during a total eclipse unless you have used charts to determine the exact location of various stars at that time. For a variety of reasons, however, there are some totalities which are much darker than others. At some eclipses, observers have glimpsed stars considerably fainter than the ones of average brightness visible on a clear, moonless *night* in the country. Surely this was accomplished by people who knew precisely where to look and who had worn dark glasses up until totality to enhance their eyes' adaptation to darkness (since it takes the eyes a while to adjust to darkness before they can begin to see the faintest objects).

Even without special measures, sometimes it is possible to see quite a few stars with ease during totality. Why is this not possible at ordinary times when a thick cloud covers the sun? Why do the stars not get brighter when a cloud hides the moon's radiance for a while? Since only a small area around us is shadowed by a cloud, when we look out in the distance we are still staring through a considerable length of brightly lit air which easily drowns out the radiance of the stars. Only the shadow of the moon is large enough to darken the atmosphere for many miles around us while still allowing us to gaze unobstructedly out to space.

Even the moon's umbra, however, is not large enough to shade all the visible atmosphere. But the fact that it is not leads to yet another utterly magnificent sky effect, the ring of deep orange light around the observer's horizon. This is light leaking to us from 30 and more miles away, from just outside the umbra where a thin crescent sun is shining. The light reaching us is predominantly orange to red because those long wavelengths are scattered far less by air than the shorter ones (such as blue) and therefore survive the long trip. The effect can sometimes be caused by a very large area of cloud covering most of the sky, but never with anything like the extent or vividness seen at total eclipses.

The beauties of the sky during totality are very great, but of course the center of attention, the focus of wonder in the whole vast spectacle, is the eclipsed sun itself. The seemingly jet-black disk of the moon is silhouetted by the pearly glimmering of the corona. This is the outermost layer of the sun, lying above the blinding (but now moon-covered) photosphere and just above the *chromosphere*. *Chromo-* means "color"; the chromosphere has received its name from its brief appearances at second and third contact, when it flashes out a beautiful pink.

Fig. 24. *A long exposure reveals the outer corona much as it appears to the naked eye. This is the February 1979 total solar eclipse as seen from Gainsborough, Saskatchewan.*

The chromosphere is only about 6,000 miles thick, but the tenuous matter of the corona ghosts millions of miles out. (It is even possible to consider the solar wind which reaches Earth an extension of the corona.) Long-exposure photographs at eclipses may reveal streamers of the corona incredibly far out from the sun, but the eye can often detect such *coronal jets* as far as several solar diameters. Within the large glow of the outer corona is the much brighter inner corona that shines in a thin band about the lunar disk. The most exquisite detail, the flowering corona, may be glimpsed throughout the corona, especially with optical aid. The shape of this flowering corona varies with the sun's cycle of activity. When the sun is near the phase of minimum activity of its usual 11-year cycle, the corona is somewhat irregular, with long streamers from the sun's equator and short, curved rays from the poles (*polar brushes*). Near solar maximum, the corona is more regular in its shape, which is often compared to that of a dahlia blossom. In any total eclipse, it may be possible to see other structures in the corona—loops and intricate twistings. No one photograph can come close to doing the corona justice, partly because only the eye in direct observation can respond to the very wide range of brightness from inner to outermost corona. (But special films and techniques are producing ever more marvelous results.) Books usually show exposures that were long enough to capture the corona flaring out to a moderate distance, but to make these photographs requires that the inner regions be overexposed. Such photographs almost give one the impression that the corona is a fierce and nearly uniform glare around the moon. Nothing could be farther from the truth about this delicate and detailed radiance.

There is another solar phenomenon visible during total eclipse that is as colorful as the briefly glimpsed chromosphere. As one layman resident of Lundar asked out loud while fixing his naked eyes on the eclipse: "What are those red things sticking out from the sun?" They are *prominences*, which look like tiny plumes or tongues of flame to us from 93 million miles away—but many of them are really geysers or fountains of flowing matter which would dwarf our Earth: some reach out to distances of a hundred thousand miles and more from the sun.

They seem motionless—frozen flames—but once again this is only due to the distance from us; some of them are actually rushing at speeds of up to 1,000 miles per second. They are not always visible to the naked eye (when the sun is inactive there may be no major prominences), but binoculars show them superbly. Their vivid color is amazing; the thought of what they are is humbling: that such hugeness and fierceness could be so pretty and delicate when seen across the great gulf of space!

All these wonders of total solar eclipses must be viewed in just a few minutes—at Lundar, 2 minutes and 50 seconds. The great African eclipse of 1973 was total for just over 7 minutes at the best location, one of the three longest totalities of the millennium, and not much short of the maximum possible (which, according to one of the ways of calculating it, is about 7½ minutes). The next total eclipse of the sun visible in North America will occur on July 11, 1991, when Hawaii will experience a 4-minute eclipse in the morning, but Mexico (including Mexico City) will have a magnificent totality of 6 minutes and 45 seconds. These still short durations are precious for their beauty but also for their knowledge, because although astronomers have succeeded in artificially blocking the sun's disk with instruments, the rare moments of totality provide by far the best opportunity for studies of the solar atmosphere. At the 1973 eclipse in Africa, a swift jet chased the umbra and managed to stay within totality for 74 minutes! Compare that to the amount of totality time experienced by even the most active of ground-based observers. There are very few members of the 1,000 Second Club, those people who have spent at least 1,000 seconds in the darkness of total eclipses of the sun.

What a feeling of longing there is at the end of totality, when the shadow rolls away, and the light of common day floods back over all. The eclipse was so brief: will you ever get to see another? But those last seconds can be the most awe-inspiring of all, for you again may see the chromosphere and Baily's Beads, and, if you are lucky, an eclipse phenomenon which I have saved for last because it is possibly the most breathtaking of all: the *diamond ring*.

If at the beginning or end of totality there is a deep valley

Fig. 25. *The bright photosphere of the sun shines through a lunar valley, creating a beautiful diamond-ring effect. Photograph by Anthony Pirera.*

properly situated on the moon's edge, then the last or first speck (far tinier and lovelier than a bead) may shine all by itself for a brief time—a diamond on the band of the inner corona (which is still visible despite the brilliance of the diamond). The term was first used in 1925, the last time a total solar eclipse occurred in New York City.

At Lundar the rear edge of the shadow was rising in the southwest sky. There were shouts of "It's coming back!"—the sunlight that is: so soon was this eclipse ending. But suddenly all our thoughts were halted, pierced. The loudest shouts and cries of all were now heard: the brightest "star" any of us had ever seen had appeared. It was the diamond. It was set on the upper right of the still-glimmering inner corona just like a jewel, but this diamond was indeed brighter than all of those of Earth.

The diamond, from the very first more intense and far more brilliant even than Venus or a full moon, brightened but still hung starlike for an astonishingly long time. Our wonder grew proportionately with each succeeding second, but it was the last few seconds that were the magnificent culmination to the eclipse and made this the most splendid moment of observing I have ever done. As the diamond brightened, two perfect rings of color came into view, centered upon it. It was a two-ringed cloud-corona, but never had I seen an example like this, for the smallness and incredible brilliance of the diamond made the two rings also the smallest, sharpest, and most intensely deep purple-red that I have ever seen. Most dramatic of all was the positioning. Since this phenomenon was caused by ice needles, the inner ring cut directly in front of the moon, with the outer, exactly twice the radius of the first, encircling all. And the band of the diamond, the inner corona, was still clearly visible and thus interlocked with the inner cloud-corona ring. There is no way to describe this sight fittingly or our reaction to it, though I listened to a tape recording afterwards: none of us were aware of anything but the supreme beauty and our feelings released in shouts of joy.

For all of totality, but especially in those last few diamonded seconds, there was a part of us that went in wonder beyond any bounds that man has said or knows. The diamond at last flared too brightly and turned into a wire-thin crescent blindingly intense. The shadow swept on. The light of common day returned. The total eclipse was over.

Eclipses of the sun and moon range from the merely interesting to the dumbfoundingly beautiful and strange. In Asia Minor in 585 B.C., the Medes and the Lydians were stopped in midbattle by a total solar eclipse which so awed them that they abandoned their hatred and ended their war. The lesser eclipses may not leave you quite breathless, but their beauty and peculiarity will enrich you. You should never miss a chance to see one. For the masterpiece of them all, the total eclipse of the sun, you will probably have to make your own opportunity. Travel quite a distance or perhaps time your vacation to fit the eclipse into your itinerary. You will not find any place more mystically beautiful than that brief land called totality.

At the Lundar eclipse I was part of a tour which included people of all kinds: everything from an apprentice baker to a photography professor to a woman who claimed to be a witch.

Though each of us had his or her private experience, we also shared an incredible adventure—working together in every possible way, planning, hoping to dodge all clouds and find our spot in the clear to meet the coming shadow. After the eclipse all of us were so stunned that for several hours we could hardly talk about what had happened. Then the words came, and looks at one another, smiles, with the knowledge that all of us had been through something very few people experience, something that shows a mystery in life that perhaps nothing else could do so well at the same time to so many very different people. In that respect, the event we witnessed was a miracle in the truest sense of the word. For wherever these people go, however faded the memory becomes in the problems and drudgeries of life, there will always be somewhere inside us all those few minutes of wonder that joined us shining forever in the diamond ring of rings.

Fig. 26. *The zodiacal light (glow through middle of photo) and the great tail of Comet Ikeya-Seki were visible to the naked eye when Dennis Milon took this 25-second exposure at 5:07 A.M. MST on October 26, 1965, from an altitude of about 7,000 feet in the Catalina Mountains near Tucson, Arizona.*

CHAPTER SIX

Comets

A vast feather of radiance hangs above the first awakening white gold of morning twilight. Small stars still twinkle all around and even through the ends of the glowing, diaphanous veil high in the sky, while down below, the big, fiery star from which the plume of light extends burns in the brightening dawn-light with intense, enduring radiance.

This is the sight of a bright naked-eye comet, which, on the average, might be seen about once every five years. At longer intervals, perhaps just a few times in a person's life, one of the truly stupendous comets appears. A great comet is the most spectacular celestial sight which most people ever get to see. In the last few centuries, there have been comets whose tails have stretched more than halfway across the heavens and whose heads appeared—and sometimes really were—larger than the sun! The brightest comets have been brighter than any star or planet; some members of the group called the sun-grazers have even greatly outshined the full moon when they were very near to the sun. Comets have struck terror into the hearts of the superstitious for ages. Even as late as 1910 a few people committed suicide for fear of the close approach of Halley's Comet. Unlike the planets, which are almost always visible and always found in the zodiac, a comet can be seen suddenly one evening, then quickly brighten, and march night after night through any part of the heavens, radically changing the appearancc of its head and especially its tail or tails. There is little wonder that the superstitious have always regarded the comets as dreadful portents. In the famous words of Calpurnia in Shakespeare's *Julius Caesar:* "When beggars die, there are no comets seen; / The heavens themselves blaze forth the death of princes."

It was Edmond Halley who about three hundred years ago began to explain the behavior of these spectacular objects and in doing so began to change men's fear to awe and admiration. The past thirty years have seen major strides in our understanding of comets, and the return of Comet Halley in 1985–1986 will provide us with an opportunity to try to learn much more, with close-up visits by several unmanned spacecraft. The twentieth century has had far fewer great comets than the nineteenth, but the prospects for upcoming years are good. Even as this book is being prepared there is a chance that astronomers will soon sight a potentially bright comet called Swift-Tuttle, and at any time a blazing comet never before seen could be discovered. The majority of comets are faint (though still fascinating) objects which must be observed with a telescope, but when a bright comet appears there is so much to see and appreciate that one still needs to be prepared. Who knows but that the greatest comet of our lives may appear tonight?

This chapter concerns the observation of bright comets. You may find yourself so inspired by the sight of a bright comet that you will decide to obtain a telescope to track the fainter members of this mighty and mysterious clan. A number of new comets continue to be discovered by amateur astronomers, and a few times in the twentieth century they have even been discovered with the naked eye. Amateur observations of comets are certainly of great value. To begin our survey of comets, we will follow the progression of a typical bright comet through its first encounter with the sun in its passage through the inner solar system.

The question of where comets originally come from is likely to remain unanswered for a long time, but we know that most of them have entered the realm of the planets on highly eccentric—that is, elongated—elliptical orbits. The far end of any of these orbits must be very much more distant from the sun than the outermost of the known planets. It is believed that comets were formed in these farthest reaches of the solar system at about the same time as the planets, and are still to be found out there in a vast *Oort cloud* (named after Jan Oort, who proposed its existence in 1950). This spherical cloud may contain many billions of dark, frozen comets, so that, just as Kepler guessed, there are perhaps "more comets in the sky then there are fishes in the sea." Many of the comets in the Oort cloud may be at distances nearly halfway to the nearest stars, far enough to have their orbital motion speeded or slowed by the gravitational influence of passing stars. It is these perturbations which could eventually cause a comet to plummet in toward the realm of the planets, and possibly far enough into the inner solar system to undergo dramatic changes which would make it spectacularly visible from Earth.

But what *is* a comet? The best answer now available is the "dirty snowball" model proposed by Fred Whipple in 1950 (a big year for the science of comets). According to Whipple, a comet far from the sun consists of a *nucleus* composed mostly of water ice (along with ices of some other substances) and an assortment of various-sized particles of silicate dust. This nucleus is probably no more than a few miles in diameter and far too faint to see by reflected light at all when it is far from the sun. If this nucleus speeds in to within about three times the Earth-sun distance, however, the increased heating causes the ices to sublime directly into gases which, along with the released silicate dust, will billow out to form a vast cloud called the *coma*. "Coma" is from a Greek word for "hair," which is appropriate for the fuzzy appearance of this dusty cloud of gas (the word "comet" itself means "long-haired"—a reference, of course, to the spectacular streaming tails). As it approaches the sun, the coma becomes ever more self-luminous as its gases fluoresce from the increasing intensity of ultraviolet light. The coma may appear more or less condensed, but, whatever its appearance, it forms with the nucleus the *head* of the comet. Some comets never develop any further than this. If, however, a large one enters as far in as the orbit of the Earth it will usually also produce the popular trademark of a comet, the magnificent *tail*.

Comas range from about 10,000 miles in diameter up to 600,000 or more. The Great Comet of 1811 had a coma far larger than the sun! At first glance, this might seem impossible (or very frightening!) but it is not if we recall that the gravity of an object depends not on its size and distance but on its *mass* and distance. The masses of all the billions of comets in the Oort cloud combined may amount to no more than a small fraction of the Earth's mass. Comas must therefore be extremely tenuous, as is proven by the fact that stars can sometimes be seen shining through their outer regions. The low mass of comets was perhaps most dramatically demonstrated by the passage of Lexell's Comet and Comet Brooks through the satellite system of Jupiter without causing the slightest disturbance. In 1886 Comet Brooks even came into contact with the outer gaseous layers of Jupiter without causing any disruption to the planet (Jupiter, however, has a very powerful influence on the orbits of many comets). When comets have passed rather close to Earth their comas have sometimes appeared as large as or larger than the moon. The same Lexell's Comet which came so near to Jupiter also passed less than $1\frac{1}{2}$ million miles from Earth in 1770, when it appeared $2\frac{2}{3}°$ in diameter, over five times the apparent diameter of the moon, both figures records. (It should always be remembered that comets have only been well studied for the past few hundred years; much huger, brighter, and longer comets must have filled Earth's skies in earlier history.) In 1976, I was fortunate enough to be one of the few people to see Comet d'Arrest with the naked eye; the close-passing cloud of the coma was a large patch of phantom light in an especially dark, star-poor region of the heavens.

Although the coma may thus appear as a sizable object, often clearly larger than a star to the naked eye, this huge cloud can itself be dwarfed by the comet's tail, which may appear spread over a large part of the heavens and extend for many millions of miles in space. The longest comet tail in actual physical dimensions (that we know of) was the tail of the Great Comet of 1843 (one of the "sun-grazers"), which extended for an almost unbelievable 200 million miles—much farther than the distance from the sun to Mars, and over twice as far as the distance from the sun to Earth! Remarkably, however, even a tail as long as this one would contain a truly minuscule amount of matter, for the tails of comets are far more tenuous than laboratory vacuums on Earth, the gas molecules being so thinly spread that one might have to travel hundreds or even thousands of miles from one molecule before coming into contact with another. There is nothing in astronomy more amazing than the fact that so little substance can produce the vastest and perhaps the most beautiful structures in the solar system.

It is common knowledge that a comet's tail always points away from the sun, preceding the head as the comet recedes from the sun. But the tail does not always point directly away. In the first place, there are actually two major types of tail which are common and which may sometimes both be observed in a bright comet. The broad and slightly reddish *dust tail* is curved and lags considerably behind the *radius vector*, the line extending through sun and comet. The explanation for the lag is that these dust particles are mostly small enough to be driven away from the head by the pressure of solar radiation, but require a considerable amount of time to travel the great length of a tail while the head of the comet is still moving onward in its orbit (you can demonstrate this to yourself by running ahead or on a curved path with a streaming garden hose).

On the other hand, the *gas tail* of a comet is composed of gas molecules which are carried swiftly back by the flow of the endless stream of atomic particles we call the solar wind. The gas tail therefore does follow the radius vector, pointing straight away from the sun, and appearing as a narrow spike or beam which is bluish (this color is usually only visible on photographs). Gas tails may exhibit streamers, kinks, and knots as a result of the magnetic-field effects of the solar wind, and irregularities in the gas production of the rotating nucleus. These beautiful silky strands of light and their twists and bundles are visible only in the telescope or in photographs, as are many wonderful details of the head. In bright comets it is sometimes possible, however, to see with the naked eye a starlike point of light in the head, which is a small central condensation about the much smaller and always hidden nucleus.

In a bright comet, that nucleus is a spinning fountain with a straight stream of magnetically stranded and twisted, fluorescently glowing gas, and a curving spray of sunlit dust, all washed and blown over millions of miles of space. But besides the constant physical changes in the head and tails of comets (some of which may be observed in hours or, with a telescope, even in minutes), there are also the changes in the appearance of the tail or tails which result from the different viewing angles we get as both we and the comet travel on. The apparent length of a comet's tail depends not only on its true physical length and our distance from it but also on the angle from which we observe the tail: a long tail might, for instance, appear extremely foreshortened if it were pointing directly towards us. Sometimes we can look upward or downward to the plane of the comet's orbit and see the dust tail as a very broad fan. Such a fan may be seen divided into several parts, so that there have been comets observed with five, six, and perhaps even seven tails! When we pass through the plane of a comet's orbit a broad dust tail appears to narrow, and sometimes it is then possible to see an *antitail* seeming to point towards the sun.

In reality this is usually an effect of perspective and what we are seeing are the dust particles too large to be driven off by solar-radiation pressure, and so accompanying the head for a while. But this may not be our last view of these particles: they may someday blaze again in the heavens as shower meteors in our atmosphere.

Magnificent tails were displayed by the finest comet of recent years, Comet West. This splendid object appeared about two years after the great expectations for Comet Kohoutek had led to disappointment, and perhaps the lack of advance publicity for Comet West was partly a reaction to the earlier event. Comet West was at its best when it was visible only in the few hours before dawn in March 1976, so it was not observed by the general public. What a loss it was to miss it! This is ample evidence that it is quite possible to miss even a spectacular comet if you do not subscribe to any astronomical publications (see the Annotated Bibliography for information on astronomy magazines and the publication *Tonight's Asteroids*). Comet West hung in the last hour of night and morning twilight with a long gas tail and a huge dust tail. On the morning of March 12, the temperature where I live was 21° F and the dust tail of Comet West could be followed for 21°. Expert comet observer John Bortle of Stormville, N.Y., observed this tail to stretch for as much as 26½°, and photographs revealed that it may have approached 50 million miles in length at best. Comet West also possessed a third, more curved ("Type III") tail and was unusual for having very prominent structure in the dust tail.

Comet West was notable for its extreme and enduring brilliance. Steve Albers was among those who observed the comet just after sunset when it was nearly closest to the sun (in space and in the sky). Since it was visible to the naked eye in this extremely bright sky it was probably considerably brighter than any of the stars. The most interesting observation of all was John Bortle's spotting of Comet West with his naked eye while the sun was above the horizon. It thus became only the third daylight comet of the twentieth century. About a week and a half later, as it began to become easily visible in morning twilight, the head of West was still as bright as any star in the Big Dipper, and its immense tails were becoming visible higher up in a darker sky. The comet continued to rise earlier each morn-

ing and surprisingly retained much of its brightness for the entire month of March. My last naked-eye sighting of Comet West was on April 10, but it would have been later if bad weather and the bright moon had not interfered. All in all, this magnificent comet was possibly visible without optical aid for over a month and a half—a time which for me was filled with a treasury of sights I could not fully describe even in a very long chapter of their own. But I remember perhaps best of all what seemed to me the liquid *gold* of the comet's head and brightest tail area, with the gas tail pointing up like the strong beam of a lantern in the sky.

There have been several other bright comets in the past few decades, though perhaps none which were both as bright and as well placed for so long as Comet West. All of these objects were members of one of the two major classes of comet: they were all *long-period* rather than short-period or *periodic* comets. Long-period comets are those which are making either their first or one of their first flights in from the Oort cloud. After one of the virgin comets is perturbed by passing stars to fall in toward the inner solar system, its immensely elongated elliptical orbit carries it back out to a far point as distant as its original position in the cloud. This new orbit therefore takes as much as thousands or millions of years to complete, and so it is only our very distant descendants who will get to see any of these comets return to shine again in the skies of Earth. It is believed, however, that eventually a long-period comet passes close enough to a large planet—usually the giant Jupiter—to have its orbit further perturbed and greatly diminished into a far less elongated ellipse, often with a far point in the vicinity of the perturbing planet's orbit. These comets thus become periodic (short-period), which by definition complete an orbit in less than 200 years. Most of the periodic comets are associated with Jupiter and so in less than a dozen years accomplish orbits with far points in the vicinity of that planet's orbit. Their point in space closest to the sun is their *perihelion* (the far point is called the *aphelion*), which may be closer in

Fig. 27. *Comet West, photographed by Betty and Dennis Milon from Sullivan, New Hampshire, on March 7, 1976, at 4 A.M. EST, using a Fujica camera with a 45-mm, f/1.8 lens. The time exposure caused the stars to trail. Comet West's 13° dust tail shows as white, its gas tail as blue (in color on front cover). The comet was very impressive to the naked eye, but it is necessary to drive away from city lights to obtain such a view of a comet.*

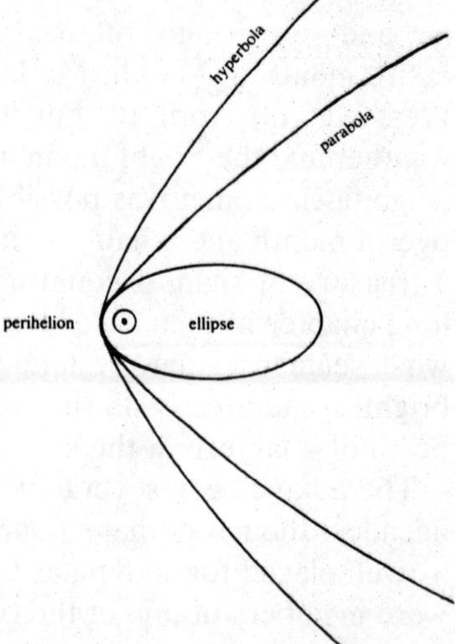

Fig. 28. *Three types of comet orbits around the sun, compared with the near-circular orbit of Earth. These particular comets happen to have their perihelion farther out than the Earth's orbit. Note that the parts of their orbits near perihelion are virtually identical and indistinguishable from one another for the three different types.*

than the orbit of Earth, Venus, or Mercury, so that we usually get to glimpse them as they pass by on each circuit, and therefore a number of times in the course of an average human lifespan.

Unfortunately for naked-eye observers, nearly all of the periodic comets are faint objects, almost always requiring a telescope to glimpse. It is not surprising that after so many encounters with the sun these objects have lost much of their former glory. The comet with shortest period whose orbit has been definitely determined is Comet Encke, which has a period of just 3.3 years and has been observed 52 times, counting the unusually fine 1980 appearance, when a few people were able to glimpse it with the naked eye. According to Guy Ottewell, meteoric evidence shows that Comet Encke has survived at least 1,500 orbitings. Will its ultimate fate be that of Biela's Comet (which we examined in our discussion of meteors)?

Perhaps. When all the gas has streamed off a periodic comet, some core of the nucleus may live on as one of the smaller of those rocky worldlets we call the asteroids. But it has long been known that Comet Encke is the source of the widely spread though often bright Taurid meteors, and it is now believed that the mysterious object which exploded over the Tunguska region of Siberia in 1908 was a fragment of the nucleus of Comet Encke. On the morning of June 30 of that year, people over a vast area of Asia observed an object which must have rivaled the sun as it crossed the sky. When the object finally exploded low over the wild, virtually uninhabited Tunguska region it produced a blast that devastated the forest over an area as much as 60 miles in diameter and knocked down men and horses in the range of 100 to 150 miles away. From a distance of about 250 miles observers saw a jet of flame rise up to a height of roughly 12 miles. On the following nights, bright noctilucent clouds were seen all the way to the Atlantic. Theories about the Tunguska object have proposed that it was a meteor or a piece of antimatter or a miniature *black hole* or even a crippled nuclear-powered spaceship, but the evidence is now very strong that this body was a piece of the nucleus of Comet Encke. If the Tunguska object had arrived 4 hours and 47 minutes earlier it would have destroyed the city of Leningrad (then Saint Petersburg, the capital of Russia), a city of about two million people, and not many years later the main setting of the Bolshevik revolution.

Although most periodic comets are too faint ever to be seen with the naked eye, one of the few exceptions is also the most famous comet of all, the illustrious Halley's Comet. Edmond Halley, one of the greatest astronomers, was not the first to see the comet which bears his name, but he was the first to realize that it was returning at intervals of about 76 years (on the average), and the first to determine the orbit of any comet. Halley and Newton were among the many interested observers of the comet in 1682, and Halley later predicted that it would return in 1758. He died in 1742, at the fine old age of 85, but he must have been confident that he was right. The year 1758 was nearly over, however, and no comet in the proper orbit had been sighted, when at last, on Christmas Day, the amateur astronomer Georg Palitzsch came across a faint stain of light in the right place. By the following spring the comet had brightened to prominence. There could no longer be any doubt that Halley's prediction had proven correct.

Checking back through the historical records, researchers have discovered 27 definite returns of Halley's Comet since the appearance of 87 B.C., and three likely earlier returns, the first having been in 467 B.C. At each appearance, Comet Halley has lost more of its material, so that its displays in recent centuries have been, on the average, considerably less bright and spectacular than those of medieval and ancient times (some astronomers, however, believe the dimming has been relatively slight). At any individual return, though, the splendor of Halley's Comet is also very strongly dependent on how closely the comet approaches Earth. The 1910 return was very impressive because the comet passed about 14 million miles from Earth, and around May 18–19 our planet may have traveled through the outer fringes of the tail. That day astronomers were in the Hawaiian Islands to watch the head of Halley's Comet *transit* (pass in front of) the solar disk, but were unable to detect any presence of the tenuous coma or tiny nucleus (the same thing happened when the great comet of 1882 transited the sun).

The astronomical positionings at the upcoming return of 1985–1986 will be far less favorable. Comet Halley will pass several times farther from us and when at nearly its best will appear very low in the south as seen from 40°N (farther north, in Canada and Britain, the head of the comet will certainly be lost from view for a while). Because of the comet's distance, it will appear, when best, as an object of only moderate naked-eye brightness. On the other hand, Halley's Comet has never been most noted for its brightness: its fame derives from its grand, long tail and the fact that it alone of the great comets returns as often as once in most human lifetimes—often enough for it to enter many a scene in history and to give each generation on this Earth a shining torch of wonder to look forward to for years and remember for the rest of their lives. In 1986, Comet Halley's tail could be comparable in length to Comet West's or possibly even much longer—it may extend as far as 20° to 40° or more in a dark star-filled sky. But there is one problem: that sky is exactly the kind which millions of people will not have—unless something is done.

Unless we act, the proper sight (or, in big cities, any sight) of Halley's Comet will be denied millions of people because of one factor: the problem of "light pollution." Light pollution is the *excessive* use of artificial lighting by cities, towns, and businesses. It not only robs us of starry skies (one of nature's greatest wonders) but also of billions of dollars and something else more important to all of us than money: its contribution to the energy crisis helps jeopardize the very stability of international relations, affecting (directly and indirectly) the quality of our lives and threatening even life itself. Perhaps the greatest gift of Halley's Comet would be if its renown could bring this critical problem of light pollution to public awareness. I propose that individuals everywhere talk to their local government (and their friends, of course) about the idea of dimming artificial lighting at special times on the best nights of Halley's Comet in the spring of 1986. If you want information on this idea and on observing Comet Halley at the coming return, you can write to the author in care of *Dark Skies for Comet Halley*, RD #2, Box 248, Millville, New Jersey 08332. As of autumn 1982 this idea has already won support from the largest professional and amateur astronomical organizations in the United States, but there will always be too little time left for what still needs to be done—the support and interest of individuals everywhere are needed as soon as possible if most people's proper sight of Comet Halley is to be saved, and the general problem of light pollution recognized.

The spirit of sharing a view of Comet Halley in dark skies would be a beautiful one. We would be looking at that special celestial visitor which has encountered our planet at once-in-a-lifetime intervals and which has borne witness to so many episodes in that long tale of joy, woe, and striving. The comet itself was witnessed with wonder or fear by the combatants in A.D. 451, when Attila was at last defeated; by Normans and Saxons in 1066; by the apocryphally comet-excommunicating Pope Calixtus III in 1456; and by Shakespeare, Galileo, and Kepler in 1607. Mark Twain entered this world with the comet in 1835 and left with it in 1910. At its next return, this great comet should at least be met with the welcome of at least a few of our unmanned spacecraft, which will have come to gaze at and learn about its wonders. Though at present it seems sadly unlikely that any of the members of that welcoming party will be from the United States, let us hope that somehow Halley's Comet will pass a world that is at least in some ways wiser and more appreciative of that great visitor's glory.

CHAPTER SEVEN

The Planets

The fact has been expressed many times by a variety of people in the past few years, but it still bears—indeed cries for—repeating: we are alive at the birth of the first great age of planetary exploration. Our generation is the first to see the "wanderers"—Greek *planetai*—as more than just points of light or tiny disks in the telescope: we have begun to see them as true worlds, as real places. We are the first entire generation to perceive clearly that Earth is really a member of a family, that the other planets are brothers and sisters both like and unlike her. In just the last decade, we have learned hundreds or thousands of times more facts about the planets than we knew previously; and, most importantly, some of the new findings have been of fundamental significance. Even the first manned visits to these worlds will probably not uncover new visions that are so broad (though they will surely be deeper), nor are those visits likely to occur within a few years of one another, in the staggeringly dense concentration of knowledge-pregnant measurements which has deluged us in the late 1970s and early 1980s. This revolution of meaningful knowledge, this revelation of beautiful vision, is probably even greater than those bequeathed to, and in part obtained by, Galileo and Kepler.

We who are alive in these times are in a very privileged position, but with our privilege inevitably comes a responsibility. That responsibility, if understood clearly, is quite the opposite of a burden. Put most simply, it is the responsibility to cherish what through the efforts of so many people we are at last able to see and even visit. The very fate of our own planet depends upon it, and I think that the first step in the right direction is in our attitude toward the view of the planets we can see with our unaided eyes. We should watch the planets not with less interest now that we can see them close up in photographs and read much about them that was previously unknown—on the contrary, we should follow them with even greater interest and appreciation now that our eyes can be filled with greater understanding. Is that not precisely the same progress we make in our relationships with all human individuals whose worth we come to know? The "mystery" of something about which we are grossly ignorant is very different from the deep mystery we find and love when we come to know something or someone truly well!

From earliest times the naked-eye planets have been known as individuals, with characters consisting of their brightness, color, speed, and behavior. In our survey of the planets in this chapter, we will look at these distinctive qualities with a greater depth of insight that is founded on many of the marvelous facts revealed by recent space explorations. I hope that this brief guide will help naked-eye observers enjoy the wanderers more than ever.

Before we tour the individual planets, there are some general points to be made about the appearance and location of these objects as a class. Many beginners wonder first of all how to distinguish a planet from a star. There is, of course, little difficulty if you follow the planets consistently enough to keep track of where they are each week or month. The planets were called wanderers by the ancient Greeks because they change their position from night to night against the background of the stars, which remain in virtually the same positions with respect to one another for thousands of years. So a planet can be detected by its motion after a period of several nights or weeks, and, of course, anyone who knows the patterns of the constellations can immediately detect an intruder. It is also true that all but one of the planets which can be glimpsed with the naked eye are bright—but only a few planets usually outshine the very brightest star. So is there any foolproof way to tell a planet from a star on a single night, save by learning the planets and constellations? One very helpful clue is whether or not the object twinkles. Twinkling is caused by atmospheric turbulence that causes deviations in the path of the light from the point image of a star, creating continual changes in the star's position, brightness, and even color. And so stars twinkle, especially when seen low in the sky. On the other hand, although planets seem like points of light to the naked eye, a telescope shows them as tiny spheres or, considered two-dimensionally against the sky, disks. Since light is coming to us from all across this small but sizable disk it is far more difficult for air currents to deviate enough of the planet's light sufficiently to make it twinkle. A planet almost always shines with a steady, serene light which is quite distinguishable from that of a star.

Another clue to the identification of an object as a planet is its location. The planets, except for strange and telescopic Pluto, are always found somewhere in the band of about a dozen constellations which we call the zodiac. The central line of the zodiac is the ecliptic, the sun's apparent path among the stars. We can never see the sun among the stars, of course, but when a constellation of the zodiac keeps setting sooner and sooner after the sun it is easy to figure out that the sun is about to enter that constellation. This apparent motion of the sun among the stars is really the motion of our observing platform, Earth. Why then do the moon and planets spend all their time traveling near to the ecliptic? Since the ecliptic is the projection of the Earth's orbit in the sky, it must be that the moon and planets travel in nearly the same plane as we do. In other words, if we could look at our entire solar system from beyond the orbit of the outermost planet, viewing it from the side (neither above nor below the Earth's orbit), we would find that the orbits of the other planets are only very slightly inclined to the plane which the Earth travels in. The only dramatic exception is the orbit of Pluto, a body so tiny that some astronomers believe it should not be considered a planet (there is even a speculative theory that Pluto was once a moon of Neptune).

The first planet one is likely to notice—can hardly, in fact, avoid—is the most brilliant, Venus. Like all planets, Venus varies somewhat in brightness (for several reasons, one of which is its varying distance from us), but even at its least radiant this planet easily outshines all others. It is little wonder that this yellow-white jewel has been the most famous celestial object after the sun and the moon and has entered so much lore and literature, usually in one of its two aspects: the Evening Star or the Morning Star.

As many people know, the Evening Star is a brilliant point of light seen in the west for a while after sunset, and the Morning Star a similarly blazing object visible in the east for a while before sunrise. Tennyson, in his great poem "Crossing the Bar," writes, "Sunset and evening star, / And one clear call for me!"; Thoreau ends *Walden* with the statement that the sun is but a morning star—that is, the sun as we now see it is but a bright precursor of a (spiritually) still brighter sun which will rise when our spiritual state has improved, when we are more "awake" to the world and ourselves. Both Evening Star and Morning Star are Venus, visible at those times when Venus (which is closer to the sun than we are) emerges on either side of the sun from our point of view.

The accompanying diagrams (Figs. 29 and 30) show positions of Venus in the evening sky during a rather good apparition (period of visibility), and the orbital positions of the two planets

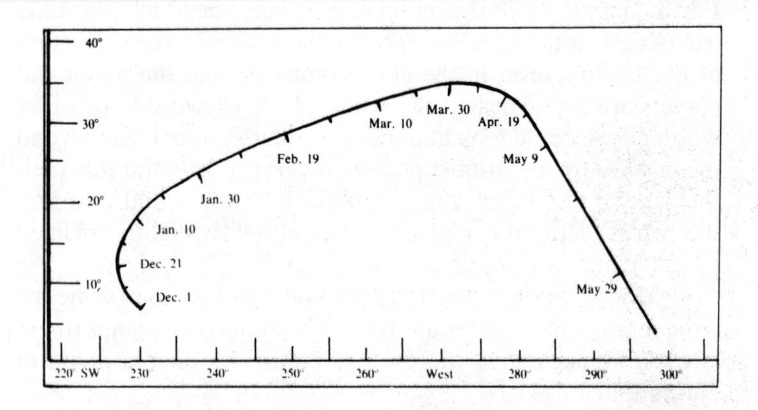

Fig. 29. *An excellent apparition of Venus in the evening sky for 40°N, showing the planet's position on various dates at 45 minutes after sunset. Vertical figures are degrees of altitude above the horizon; horizontal figures are degrees of azimuth around the sky. Notice how quickly Venus changes position as it nears Earth, approaching an inferior conjunction occurring on June 15; greatest elongation is on April 5 and greatest brilliancy on May 9 in this particular instance, which will be nearly duplicated in 1988.*

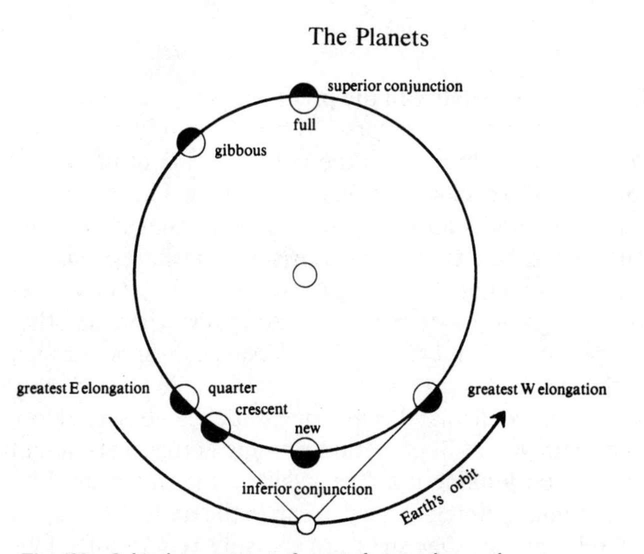

Fig. 30. *Orbital positions of an inferior planet (here, Venus). Compare with graph of the apparition of Venus (Fig. 29).*

which account for what is observed. The most favorable position for observers is clearly *greatest elongation*, when Venus attains its greatest separation from the sun in the sky—as much as 47°—so that it may set as much as four hours (and slightly more) after the sun. Also of major interest is *greatest brilliancy*. This does not occur when Venus is closest to us, because at that time (inferior conjunction with the sun) we are viewing the night side of the planet, as the orbital diagram shows. Greatest brilliancy always occurs when Venus is about 39° from the sun—a position which provides us with the best combination of least distance from Earth and greatest illumination of Venus's Earthward hemisphere. Just before *inferior conjunction*, when Venus passes almost between us and the sun, and is as little as 25 million miles away (the closest any planet can approach), the illuminated crescent is skinny but huge—so large that it can actually be glimpsed in binoculars. There is a slight possibility that someone could then just barely detect the crescent shape of Venus with the naked eye, but there is no certain record of this feat ever having been accomplished. After the close and therefore apparently fast-moving Venus dips rapidly down to set with the sun at inferior conjunction, it climbs in just a few days to visibility in the predawn sky, where all these positions are repeated in reverse, returning the planet to *superior conjunction*, directly on the opposite side of the sun.

Although these special positions are highlights in the eternal progression, Venus is an impressive sight whenever it is visible. Part of the beauty of Venus arises from its association with all the phenomena of twilight, and from its ability—because of its brightness—to be seen in situations where all other heavenly points of light are invisible. Venus can shine through rather thick clouds and can actually cast shadows in a very dark setting. I recall that I was once shocked to see a very prominent pathway of illumination running across a stretch of glazed snow; on looking up, I realized that it was caused by the light of Venus shining through a gap in the trees! In contrast with the sublimity of that sight are the silly reports of UFOs that turned out to be Venus, and the times in World War II when Venus was actually fired on by ships who thought it was an enemy aircraft—how embarrassed they must have been to find that their shots had fallen short by a matter of at least 25 million miles!

Venus is so bright that it can often be glimpsed in broad daylight by somebody who knows exactly where to look. A good way to locate this tiny point in all that big daytime sky is to keep following it back earlier each evening in relation to objects in your landscape; the best way of all is to keep it in sight after sunrise, when it is the Morning Star. The latter type of observation is exactly what Shelley is accurately describing when he writes in "To a Skylark" of the keen "arrows / Of that silver sphere, / Whose intense lamp narrows / In the white dawn clear / Until we hardly see—we feel that it is there." When it is near greatest brilliancy Venus is especially easy to see in the daytime. There is even a story that Emperor Napoleon once walked out in broad day to find a crowd of his subjects facing away from His Majesty, staring at the sky; he soon discovered that they were all marveling at the sight of Venus in the blue. It is occasionally possible to see Jupiter and Mars while the sun is low in the sky, but to see either (or less-bright objects) in the middle of the day is a rare feat that may require particularly good vision. The most important factors in making many such observations, however, are really interest and effort: you may amaze some astronomical authorities with how soon after—or even before—sunset you can see a particular bright star or planet. Most textbooks are very conservative in their figures for all such observational tests. Few people have the enthusiasm to bother with such feats, but besides the fun of doing something incredibly rare—maybe even unique—there is the beauty of the observed object itself. Anyone who comes to admire a planet like Jupiter will want to see it with the naked eye in the daytime and enter into the small fellowship of people who have had this experience. How marvelous it is to see Venus as a surprisingly prominent point of light in the blue sky, together with the sun it orbits!

The main reason Venus is so bright is the fact that it is eternally shrouded in highly reflective clouds. Those clouds have helped keep Venus a great mystery, but a few spacecraft have managed to begin unveiling the secrets of the planet named for the goddess of love. What they have found in the atmosphere and on the surface of Venus, however, is anything but gentle or lovely. The reflective clouds of Venus form a layer stretching from roughly 20 to 48 miles above the surface, and there no longer can be much doubt that they consist largely of sulfuric acid. The atmosphere of Venus is so dense that the barometric pressure at the surface is about 90 times that of Earth, and the chiefly carbon dioxide composition of all that atmosphere has greatly increased the temperature of the planet. Venus orbits at a distance from the sun of about 0.72 *Astronomical Units* or A.U.s—that is, 0.72 of the average Earth-to-sun distance. That is about twice as far from the sun as the average distance of Mercury, the innermost planet, but Venus turns out to be even hotter than Mercury because the "greenhouse effect" of Venus's atmosphere allows little heat to escape back out to space. The temperature on Venus can be as high as 900° F! The few spacecraft which have reached the surface have survived the formidable conditions for only a short time, but long enough to photograph a rocky ground and show that the illumination at the surface is roughly that of a cloudy day on Earth. Refraction in the dense atmosphere must create truly bizarre distortions of the view on Venus. Radar measurements show that the planet has rugged topography in many places, including a great plateau ringed by mountains whose tops are as high as six miles above the surrounding plain.

Venus was once thought to be a twin of Earth: no two planets approach each other more closely, and the diameter of Venus is about 7,610 miles—only about 300 miles less than Earth's. But Venus is a very strange planet indeed. It is the only planet which rotates from east to west, and all the predictions for its rotation rate were wrong. Those estimates for the length of the Venusian sidereal day (the amount of time it takes Venus to spin around once in relation to the stars) ranged from far less than 24 hours to 224⅔ days—the length of Venus's own sidereal year. Instead, radar has revealed that no estimate within even that vast range is correct: the backward-turning Venus accomplishes one complete rotation in just over 243 days. The day on Venus is longer than the year!

So Venus and Earth are very different worlds—at present. At present, Venus is one of the most unpleasant places imaginable for human beings; but might it someday be "terraformed"—that is, changed to resemble Earth? It would be an undertaking far beyond our present capabilities, but it is not inconceivable. The question is whether it is moral. Certainly if there is any native life which has somehow managed to survive on Venus, we would be committing an act of very questionable morality to terraform the planet. But could we ever ascertain that Venus was absolutely lifeless? And even it it is a dead world, could we even then justify the act? The terraforming of any planet is a deed which carries with it the very heaviest responsibility, and we should certainly never proceed on such a course until we can honestly convince ourselves that it is a wise one. Wisdom has been a quality often notably lacking in much of our society, or at least in its leaders. There is no better proof of this than the antiterraforming of our planet which continues to proceed apace: we may be turning Earth into a Venus. Carl Sagan is one scientist who has eloquently discussed this grim possibility, which could become a certainty if we do not more wisely control our technology. Already we may have irreparably damaged some parts of the life and health of Earth. Fortunately, it seems that so far the rise in temperature caused by our increased release of carbon dioxide has been offset by the cooling effect of cloudier skies caused by pollution. But recent studies indicate that this stalemate may be changing to the warmer. We cannot be sure—and what we are risking is the very life of our planet.

Some may find it disillusioning that such an object of beauty in our skies is a world of corrosive, crushing, electrifying, frying, and poisonous atmosphere. But from a scientific viewpoint the conditions on Venus are fascinating, and if they can serve as a warning to our planet then they may be valuable indeed. Furthermore, for those of us who love the splendor of the Evening Star beckoning us like a queen to the mystic realms of twilight, the irony also works the other way: when seen from afar, even a world as frightfully inhospitable as Venus is an inspiration, a flame to our sense of beauty. Let us keep our own planet a place where such beauty can make itself manifest in clear evenings and dawns.

A world even more renowned than Venus in the legends of modern times is the Red Planet, Mars. Contrary to what we might expect, it is usually not the second most conspicuous planet, but its orbit is quite elliptical so that in certain years we can pass almost as close to this, our outer neighbor, as we can to Venus, our inner neighbor. The very best encounters of Earth and Mars occur only at intervals of 15 or 17 years, but what happens then is a marvel to behold: in just a few months we watch a planet no brighter than many stars increase immensely in radiance, becoming at last even brighter than Jupiter. Only Venus outshines Mars when the Red Planet reaches its greatest brightness, but Venus, beautiful in twilight, is also a prisoner of twilight: as a planet closer to the sun than Earth, Venus can never depart far from the sun from our point of view, and therefore can never be seen very high in a fully dark sky.

Mars, on the other hand, travels in an orbit just outside the Earth's and as an outer planet its behavior is quite different from that of Venus. From week to week, Mars creeps along against the background of stars, moving slowly eastward among them. At the beginning of an apparition of the planet, it appears in the predawn skies with the stars of a certain season and, like those stars, keeps rising longer before the sun as the weeks pass. Its eastward motion in its orbit works against this westward (earlier rising) trend, but Mars cannot completely overcome the impetus because the latter is caused by the Earth's orbital motion, and the Earth is faster than outer planets. Eventually Mars (or any outer planet) is rising at midnight, brightening all the while because when it rose just before dawn it was on the other side of its orbit, near superior conjunction (an outer planet cannot pass between us and the sun and so cannot come to inferior conjunction), but now is starting to be overtaken by the faster Earth. When Mars finally rises

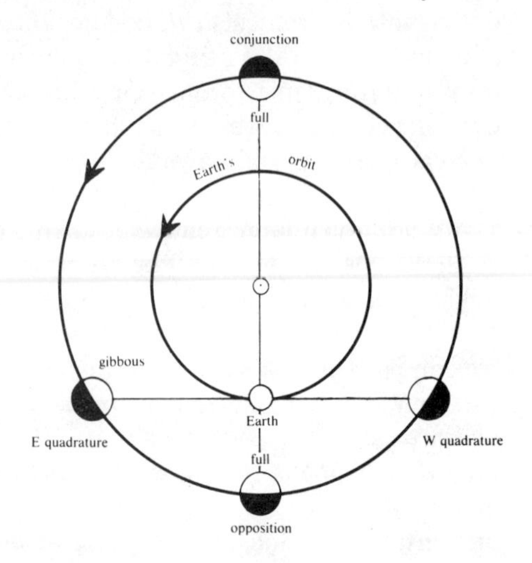

Fig. 31. *Orbital positions of a superior planet. At oppositions, the planet is closest, brightest, and visible all night long.*

at sunset it is at *opposition:* opposite the sun, visible all night, and (as our diagram shows) closest and therefore both brightest and (in telescope) biggest.

Opposition is thus by far the most favorable time to observe an outer planet (the inner planets Mercury and Venus cannot come to opposition), and in the case of Mars some oppositions are dramatically better than others. We draw even with Mars in the orbital race roughly once every 780 days and see it at opposition, but some of these times Mars is out near aphelion (its orbit is markedly eccentric), so the separation between the two planets is still rather large, and Mars is not quite as bright as the most brilliant star. After an aphelic opposition, however, the oppositions keep getting better each 780 days until our passage of Mars is occurring when Mars is near perihelion. At the perihelic opposition, which occurs once every 15 or 17 years, Mars is an imposing ruddy beacon, shining brighter than any other point of light in the midnight sky. It is not surprising that certain cultures have identified this planet with their fearsome god of war.

After opposition, the dimming Mars continues to rise earlier, so that it is seen farther west each night as darkness falls—and finally it is setting just before the sun, disappearing in the evening twilight, where it will eventually reach superior conjunction, to start the full cycle all over again. Like all outer planets, Mars appears to display a very peculiar motion when it is near opposition, a behavior which perplexed ancient observers. The latter generally assumed that the planets and sun all circled Earth, and were therefore hard pressed to explain this peculiar *retrograde motion*. At some point, a while before opposition, an outer planet will appear to cease its usual slow eastward movement against the background of the stars; it reaches a "stationary point" and appears to start moving backward—retrograding—among the stars. At a certain time after opposition, the planet will likewise halt its westward (retrograde) motion and begin moving eastward with its customary *direct motion* again.

What is the explanation? The outer planet does not, of course, really stop and reverse its motion in its orbit. What causes the appearance is the earth's overtaking of the slower planet, which shifts our viewing angle rapidly. The retrograde loop which an outer planet describes against the starry background is larger the closer the planet is (so Mars has the biggest retrograde loop); the amount of time which an outer planet retrogrades is longer the farther the planet is (so Mars retrogrades more briefly than Jupiter, Saturn, and still more distant worlds).

The striking brightenings of Mars at long intervals, and its red (really deep orange) light had a powerful impact on the minds of certain cultures, but in the twentieth century the planet rose to new heights of fame as a subject of science fiction. The original source for the tales of (usually invading) Martians was the stir created by the astronomer Percival Lowell, who around the beginning of the century thought that he could sometimes observe in the telescope the narrow lines on Mars which the Italian astronomer Schiaparelli had called *canali*—Italian for "channels," but easily mistranslated into English as "canals." Canals, of course, are artificial structures, and Lowell speculated that Mars was inhabited by an advanced race of beings who were endeavoring with great works of technology to prevent their planet from dying.

The two Viking spacecraft which reached Mars in 1976 closed the book on these stories, but the beautiful and marvel-filled wilderness which they revealed is far more exciting—and possibly not devoid of life. The *canali* turned out to be optical illusions (telescopic observation of elusive fine details is very difficult). But Mars proved to be a world of planetwide dust storms, frosts, and mists, but most of all of red sands, the solar system's biggest volcanoes, and the Valles Marineris—a canyon several times deeper than Earth's Grand Canyon, 3,100 miles long, and as much as 100 miles wide. The volcano Olympus Mons (Mount Olympus) is about 375 miles wide across its base, has a crater about 40 miles wide, and towers about 16 miles above the surrounding plain—over $2\frac{1}{2}$ times the height of Everest! The fine dust in the atmosphere of Mars makes it a world of pink skies and blue sunsets. But is it a world that supports life? The results from the Viking craft were ambiguous, but the possibility certainly exists, even though the atmosphere is extremely scanty and water is scarce (the ice caps, which are partly carbon dioxide ice and partly water ice, are very thin, though extensive). Whereas the atmospheric pressure on Venus is almost a hundred times greater than that on Earth, on Mars it is somewhat less than one hundredth of that on Earth. Carbon dioxide is the major component of the Martian air, forming—just as it does on Venus—about 97 percent of the atmosphere. At least the temperatures on Mars are

Fig. 32. *Photographed during Viking 2's approach, this dramatic picture of Mars was taken August 5, 1976, from a distance of 260,355 miles. Viking 2 approached Mars more from the dark side than had Viking 1 in mid-June, providing a crescent view of the planet. Photo by courtesy of JPL, NASA.*

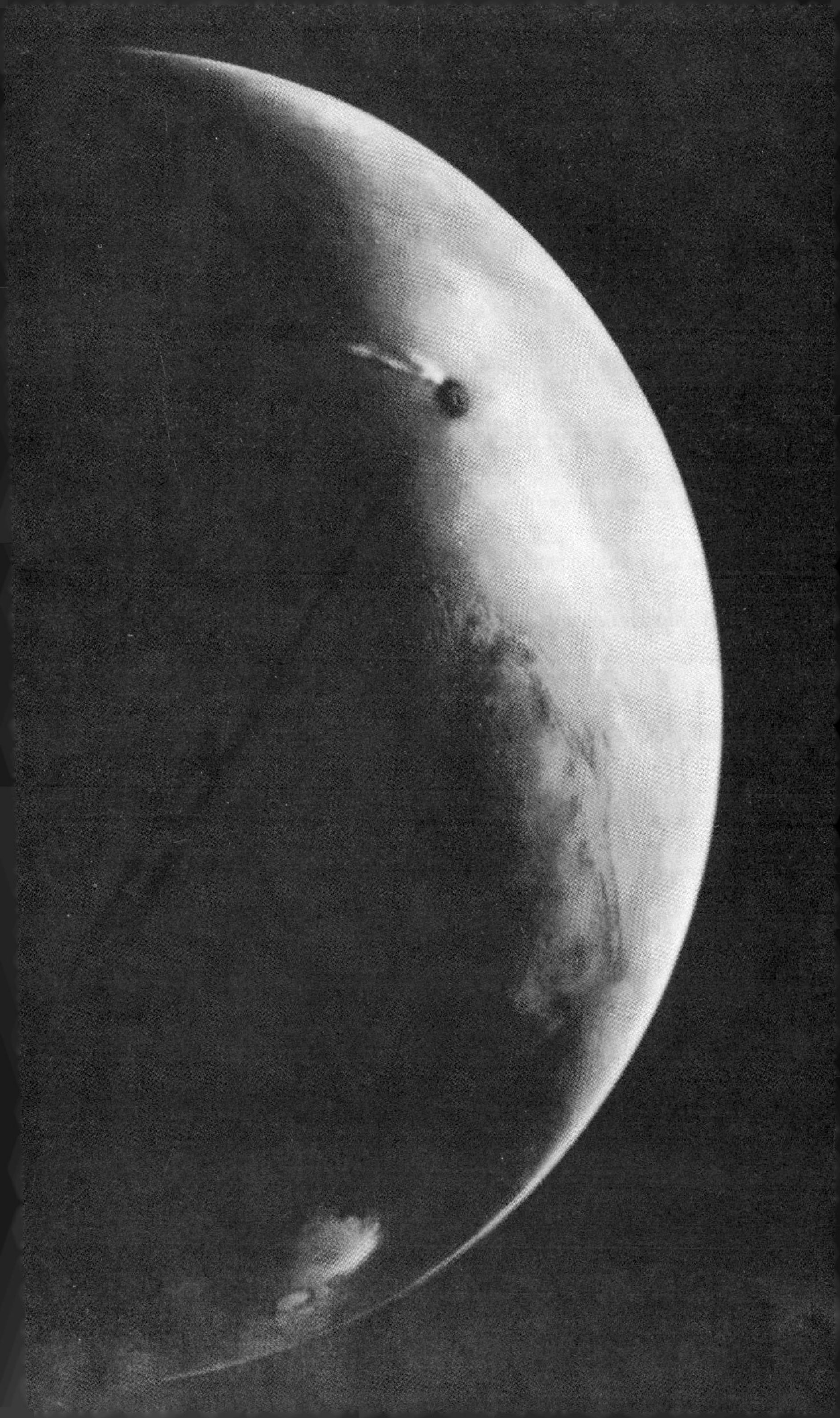

endurable: although it is usually as cold as or much colder than Antarctica, the warmer parts of Mars can, on a "summer" day, be well above freezing.

It would not be fair to leave Mars without a look at its tiny, irregularly shaped moons. These are too dim to be glimpsed from Earth with anything less than a fairly large telescope, but our spacecraft have given us some close views. Phobos and Deimos are, respectively, about 12.5 × 14.5 × 17.5 and 6 × 7.5 × 10 miles in diameter, and orbit extremely close to Mars. Phobos is so close that it will eventually crash; so close that it outruns the rotation of Mars and would appear to rise in the west and set in the east, and sometimes also rise in the west again before the night is through; so close that its equator-hugging orbit would prevent it from ever being seen at Martian latitudes higher than 69°! And the farther-out Deimos is so small that it would appear as the smallest shape resolvable to average human vision—or perhaps to some people as merely a very bright point of light. A Martian day (one rotation of the planet) is about a half hour longer than our 24-hour day and, since Deimos takes not much longer than this to complete one orbit of Mars, it would require several full Martian days to creep from one horizon to another in the Martian sky. An observer on the surface of Mars would be confronted with two little, curiously shaped (and hugely cratered) moons, one astonishingly fast, the other fascinatingly slow.

The surface area of Mars is about equal to the land area of Earth. When we finally come to visit it, this magnificent wilderness will provide people of the next generation (and perhaps some people of the present one) with many a year of exploration. And Mars, which really does have dry river beds that in past ages surely spoke with the flow of great currents of water—Mars will again be touched with sounds. The Red Planet, which has for so long been powerful in the mind of man, will now become a home—if only temporarily—for visitors from the Blue Planet.

Far out beyond Mars, past the vast region we call the "asteroid belt," lies the giant of the sun's family: mighty Jupiter. This one world is over 11 times as wide as Earth, and contains more matter than all of the other planets combined. To the naked eye here on Earth, Jupiter is the most dependable of bright planets, visible about 11 months a year and usually outshining all points of light in the sky except Venus. Jupiter is so much slower than Earth that our planet can make one full circuit and then require just one more month to draw even with Jupiter again. In other words, the planet's opposition occurs once every 13 months (a month later each year). The giant takes almost 12 years to make one orbit of the sun; that is interesting because it means Jupiter spends about one year in each of the 12 famous constellations of the zodiac. During the course of the dozen years, the perihelic opposition is quite noticeably, but not spectacularly, better than the aphelic one for naked-eye observers.

Fig. 33. *Jupiter and its four planet-sized moons, called the Galilean satellites, were photographed in early March 1979 by Voyager 1 and assembled into this composite picture. The satellites are not to scale but are in their relative positions. Io (upper left) is nearest Jupiter; next are Europa (center), Ganymede, and Callisto (lower right). Photo by courtesy of JPL, NASA.*

Jupiter is awesome in almost all of its proportions. It is now known to possess 16 or 17 moons, and more will surely be discovered. Only four moons out of this throng, however, are big: the Galilean satellites. Galileo saw the four as little stars beside the disk of the planet when he turned his telescope there—and the dance of these Jovian moons about the planet is still a favorite subject for telescopes of all sizes. Even binoculars can reveal them when they are well out from Jupiter, but many books claim that none of them is ever visible to the naked eye. More recently, a few reliable naked-eye sightings have suggested not merely that this feat is possible but also that it is perhaps not extremely difficult, as long as astronomical and meteorological conditions are favorable. I belong to this school of thought with rather good reason: I have seen several of the moons of Jupiter with my naked eyes!

The two innermost of the Galilean satellites may indeed be too close to the planet to detect without optical aid, but on one occasion I could see the moon Callisto intermittently, and on another occasion I could see Ganymede rather easily. I am well aware that my resolution—that is, my ability to distinguish fine detail or separate two very close-together objects—is certainly not better than average. When Ganymede and Callisto are near their greatest elongations from Jupiter, moon and planet are easily far enough apart for the unaided eye to see them as separate objects. It is also true that when Jupiter is near opposition both Ganymede and Callisto are considerably brighter than the very faintest stars visible on a dark country night. What then is the problem in trying to spot them with the naked eye? The problem is simply that these moons are near an object as bright as Jupiter: Ganymede and Callisto can appear far out beyond the "rays" of Jupiter, but the planet makes such an impact on the eye that it is usually not possible to glimpse the tremendously less conspicuous moons. My advice is to look for a night within a few months of Jupiter's oppositions when fairly faint stars are visible but the skies are a little hazy—as evidenced by the lessened luster, or shorter "rays," of Jupiter. If such a night happens to be one on which either Ganymede or Callisto is well out from the planet, then you may very well see one of these moons with your unaided vision. If you are rather confident that you are glimpsing something near the planet, use binoculars or check an astronomical almanac or magazine to see if there really is a moon in the proper position (to make sure, you should also check a good star map to ascertain that there is no star which Jupiter just happens to be near that night). Seeing a moon of Jupiter with the naked eye may be an extremely rare feat, but I firmly believe that making this observation requires just average eyesight, a little luck, and the enthusiasm to go out and look at Jupiter on many nights. That enthusiasm is not hard to come by once you have seen how beautiful Jupiter is when it dominates the entire sky on a clear, starry evening.

Jupiter is an immense bulk consisting mostly of hydrogen and helium. These are the most common elements in the universe, but they are so light in their free forms that they have escaped from the lesser gravity and greater warmth of Earth. Most of the earth's warmth comes from the sun, of course, but both Jupiter and Saturn are so massive that their incredibly pressurized centers emit more heat than these planets receive from the distant sun. Jupiter and Saturn are, in fact, objects that are almost in the mass range which would initiate nuclear reactions and make them tiny suns themselves. These worlds in large part are in a truly gaseous state and all we can ever see through telescopes or from passing spacecraft are the outer clouds of these gases.

The vision of Jupiter and its moons revealed by the two Voyager spacecraft in 1979 was probably the most stunning and definitely the most colorful that had yet been received from space. Jupiter, with a true rotational period of just 9 hours and 50 minutes at its equator, was shown to be a globe of not just horizontal belts and zones, but of amazing whirls and many colors (even blue). The dark side of Jupiter was lit with the necklaces of auroras and sparks of superlightning, while on the day side the famous oval Great Red Spot was seen in intricate detail (this structure may be a storm—a storm many times larger than Earth and possibly ages old!). The Voyagers also discovered that Jupiter has a sparse and variable ring, quite unlike the mighty structures of Saturn. But the grandeurs of Jupiter's turbulent and colorful face were at least matched by the surprises of the Galilean moons. The moon Io is truly like nothing that has ever before been seen in space. Looking something like a red, black, and white pizza, Io is bigger than Earth's moon. The Voyagers caught its almost ceaselessly active supervolcanos erupting sulphur and hydrogen sometimes beyond the satellite's grasp to help form a strange torus which is ionized by Jupiter's strong magnetosphere. The moon Europa has strange cracks on its face. Its surface may be miles-thick ice, under which a water sea may be kept fluid by powerful tidal effects of Jupiter and the other moons (those effects are literally turning Io inside out!). There is even a theory that some kind of life could be present in such an ocean. Ganymede is the biggest moon that has been measured in our solar system (somewhat larger than the planet Mercury) and has crustal plates suggesting tectonic activity resembling Earth's. Red Callisto, almost as large, is so heavily cratered that it looks like a billion-faceted gem.

The next planet after Jupiter is the most beautiful of all seen through a telescope: Saturn, whose rings are the most magnificent and unusual structures in the solar system. These rings have probably never been detected with the naked eye, though Saturn may appear as a slightly elongated point of light at a very low magnification. To the unaided vision this planet appears about as bright as one of the brightest stars and is a somber gold in hue. This "oldest of the old sheep" (a Mesopotamian name for the planet) requires 29½ years to make one orbit or one full circling of the zodiac, so its motion from month to month is very slow. While its rings remain tantalizingly beyond the range of naked-eye resolution, they can add much to the total brightness of the point of light that one sees. The rings are presented to us at different angles during the course of the 29½-year orbit, and sometimes we see much of the broad northern or southern face of the rings. Saturn is much brighter at oppositions when a face of the rings is visible. But twice during each orbit the rings are presented to us in a side view—edgewise—and turn out to be so thin that they disappear from view for a while in all but the largest telescopes, at which times the point of light we see with our naked eye is considerably

Fig. 34. *Saturn and its six largest moons are grouped in this artist's collage assembled from images taken by Voyager 1 during its Saturn encounter in November 1980. Clockwise, starting from the far right, the satellites are Tethys and cratered Mimas (in front of the planet), Enceladus (in front of the rings), Dione (in the left forefront), Rhea (off the left edge of the rings), and Titan (at top). Photo by courtesy of JPL, NASA.*

dimmer. The main ring system visible in small telescopes extends across a space of about 170,000 miles (the globe is about 74,200 miles wide at its equator), but Voyager 2 found that in most places these rings are no more than about 500 yards thick! The rings are thus far thinner in proportion to their breadth than a razor blade is.

If the Voyagers' view of Jupiter was grand, the sight of the Saturn system from those spacecraft was glorious, showing it not so colorful as Jupiter, but even more surprising and intricate. The ball of Saturn itself, like an oblate butterscotch or almond eye, has winds as fast as 1,000 miles per hour in its clouds, which show less detail than the warmer and less haze-obscured clouds of Jupiter. Several major new rings were discovered or verified by the Voyagers and their Pioneer predecessor, but the most marvelous Voyager sight was the thousands of ringlets making up these major structures. It has long been known that the rings could not be continuous, that they must be composed of countless separate particles, but the number and complexity of the ringlets was astonishing. Strange dark "spokes" in the B ring and thin, twisted, braided ringlets composing another ring were startling—as were the tiny "sheepdog" moons which seem to herd one ring in place, and two inner moons which apparently come so close to each other that they keep trading orbits (these two are believed to be the giant halves of what was once a single moon)! Saturn's host of at least 17, and maybe more than 20, moons (besides the countless moonlets composing the rings) includes others which do share the same orbit (always chasing each other), a moon with one side many times brighter than the other, and a moon staring at Saturn like an eyeball because of a crater, one-third of its diameter, on the eternally Saturn-facing side (the impact which caused the crater was almost enough to crack this 240-mile-wide moon in two). And there is Saturn's appropriately named moon Titan, with an atmosphere thicker than Earth's, which makes it as strange as any world we know.

But what are the stupendous rings composed of? Each major ring is of a different density and probably consists of a different distribution of particle sizes. The particles themselves are countless pieces of ice but are not white; damage to the crystalline structure of the ice caused by energetic particles in Saturn's magnetosphere has tinged the ring in subtle hues of red, tan, and brown, with some blue in the densely particled and broad B ring. The particles in the A ring (outermost visible in small telescopes) and the narrow little F ring (just outside A) are probably quite small, but in the C or "crepe ring" the average size of the ice chunks is probably about one yard, with some much tinier and some much larger. The average separation between the ice "jewels" in the translucent C ring seems to be a little more than 20 yards. The ringlets of the B ring are as numerous and intricate as grooves in a phonograph record. There are major separations and concentrations in the ring system that are caused by gravitational influences of Saturn's known moons, and after Voyager 1 scientists thought there might well be numerous additional small moons "hiding" amidst the bright throngs of the rings to create the ringlets. Voyager 2, however, took a closer look which revealed no such moons in the rings, and so this most basic fact—the stunningly intricate structure of the rings—remains an unexplained mystery. The different rings appear variously dark, bright, or translucent in degrees, according to the angle from which the Voyagers viewed them (the C ring is like a stained-glass window in the cathedral of the full ring system). Even as seen through small telescopes on Earth, the rings cast shadows on the planet and are themselves shadowed in turn. The rings of Saturn are a monumental work of "natural art": the ring system consists of great complexity which is organized on all levels into beautiful and dynamic (i.e., moving) regularities—with just enough irregularities or asymmetries (color differences and a few lopsided, braided, unequally bright, and spoked ringlets) to place the whole system ultimately beyond even the cleverest simplistic classification that could make it too predictable and dull.

In our tour of the planets we have so far been moving steadily outward, but we must reverse our course if we wish to come to the least frequently observed of the so-called "classical planets" (those recognized at least from earliest historic times). Of the famous five, this one can sometimes be the second brightest, outshining all stars, and can often be the planet closest to Earth—and yet it has been seen by remarkably few people who are not amateur or professional astronomers! The solution to this riddle is the fact that Mercury, the innermost planet of the solar system, can never depart more than 28° from the sun as seen from our point of view. Mercury can never be seen more than roughly two hours before or after the sun, so that this lesser morning or evening star is not well known. This beautiful, and sometimes orange, planet can be amazingly prominent near greatest elongation at its few best apparitions of the year (there are many separate apparitions of Mercury in a year because the swift planet races back and forth between evening and morning sides of the sun). To know when one of these favorable periods of just a few weeks will occur, you should consult an astronomy magazine or almanac (see the

Annotated Bibliography), but observers at midnorthern latitudes may follow the general rule that Mercury is best seen as an evening object sometime in late winter or early spring, and as a morning object in late summer or early autumn.

Mercury is not only the most elusive of the classical planets, it is also the smallest of the five, with an equatorial diameter of only about 3,000 miles (compared to Earth's 7,927). Mercury is, of course, the fastest planet, changing its position very rapidly from night to night. The three passes of Mercury by Mariner 10 in 1974 and 1975 provided our only close-up views of Mercury to date. Only one hemisphere was photographed on those occasions, so about half of the planet's surface remains a mystery. The half we have seen well is very moonlike, but it differs from the moon in several important ways, and there are several major puzzles associated with the planet. Mercury rotates in 58.65 days, exactly $\frac{2}{3}$ of the length of its year (which is only about 88 Earth days) and this is not coincidental, but a case of gravitational locking by the powerful nearby sun. From a viewpoint on Mercury's surface (through nary a trace of atmosphere), the sun would sometimes appear three times its width as seen from Earth, and the combination of Mercurian day and year would send it through a strange ballet in the sky.

Mercury is the hummingbird of the planets, zooming and hovering close to the sun-flower, and usually lost within that blossoming light.

Are there any more planets we can see with the naked eye? No more of the classical ones. There are the "minor planets" or asteroids, which by the thousands wander in orbits sometimes far more elliptical and, in some cases, far more inclined to the plane of the ecliptic than the planets. Most of the asteroids spend all their time somewhere between the orbits of Jupiter and Mars, but a few have aphelion as far from the sun as Saturn or, in one case, much farther, and some have their perihelion as near to the sun as Mercury or, in one case, much nearer. No one knows how they formed—possibly from the destruction or the failed birth of a planet—but only two of them, Vesta and Ceres, are likely to be glimpsed with the naked eye by even a very ambitious observer in the course of his life (it is worthwhile to try, however, because it is a wonderful thing to be able to look with unaided vision at a world just a few hundred miles in diameter which is several hundred million miles away!). The brightest asteroid, Vesta, can become rather easily visible to the naked eye on unusual occasions, but there is a major planet which we have not yet discussed which is about that bright all the time.

The planet is Uranus. You will need to observe rather far from big city lights in order to see it with the naked eye, but it is often easily visible from truly rural regions. Uranus was never knowingly observed as a planet (as far as we know) until 1781, when William Herschel saw that it showed as a tiny disk in his small homemade telescope. Uranus is about four times as wide as Earth, but it is roughly 1.8 billion miles (over 19 A.U.s) from the sun, so it appears neither big nor bright from Earth. Prior to Herschel's sighting, its status as a planet had never been suspected, because it is so slow that the motion of such a dim naked-eye object was never noticed: it takes Uranus 84 years to circle the sun, thus requiring about 7 years to cross just one average constellation of the zodiac. The oddest thing about Uranus is the fact that its axis is nearly in the plane of its orbit; in other words, one might say that it does not spin more or less upright, it rolls around its orbit, with all five of its known moons circling along with the strange angle of rotation. You must refer to an astronomy magazine or almanac to find exactly where Uranus is located when you wish to observe it. When you see Uranus without optical aid you are looking at the most distant world your naked eyes will ever see.

The still farther planets Neptune and Pluto require, respectively, binoculars and at least a 6-inch telescope to observe under very favorable conditions. What primarily interests us about them here is that they are so distant. How much bigger is this solar system than our naked eyes can show us? Pluto can sometimes go as far as 4.57 billion miles from the sun, requiring 248 years to complete just one of its vast orbits. Guy Ottewell proposes a "1,000-yard model" of the solar system (in the first issue of his magazine *In Defense of Variety*); in it a sun slightly smaller than a basketball is orbited by a pinhead-sized Pluto 1,000 yards away! (By the way, on this scale the

8-inch-diameter sun is about 25 yards from an Earth the size of a peppercorn—0.08 of an inch.) Such distances are close to the very verge of picturability for us even with such a helpful model—and yet they are very small compared to the distances to the stars. The leap to the stars must be measured in the unit of distance called the *light-year,* multiples of the distance that the fastest thing known, light, can travel in a year. Light speeds at a little more than 186,000 miles (or almost exactly 300,000 kilometers) per second, and so crosses the distance to the moon in less than 1½ seconds; to the sun in about 8 minutes; and to Pluto in about 4 to 6 hours. But the nearest star is 4.3 light-*years* distant from Earth! On the scale of Ottewell's model, the farthest planet is 1,000 yards from the sun, but the nearest star, represented by another small ball of some sort, would have to be placed 4,300 miles away! Many of the stars we see in our sky are glorious suns much farther still. We are fortunate that our unaided eyes and our imaginations will easily be able to cover such immense distances when we go on to discover the stars in the next chapters.

Before we depart on that journey, there is one more topic to consider in relation to the planets. We have seen that each of the wandering points of light is a brother or sister world, and a true individual. But what could be more rewarding and magnificent than the sight of two or more of these lovely planets positioned close to each other or the moon? In the course of a typical year, the most thrilling of all celestial sights is often a *conjunction.*

In the older, looser sense of the word, a conjunction is simply a close pairing or grouping of two or more celestial objects, but technically speaking it is the passage of one celestial object due north or south of another. The moment when one object is due north or south of the other is not necessarily the moment of their closest encounter, for which the term "appulse" is often used. When one celestial body passes directly in front of another the event may be called an *occultation* ("to occult" means "to hide"), but the moon (which is almost always the occulting object) does not pass in front of a bright star or planet very frequently for an observer at any particular geographical

Fig. 35. *Occultation of Venus on the morning of July 17, 1974. Photographed by Dennis Milon on a 4 × 5 camera with a 270-mm lens at f/5.6. The High Speed Ektachrome 120 film was developed for ASA 320.*

location. A lunar occultation of a bright star or planet might be visible as much as a few times a year, but even then the moon's light is often too overwhelmingly bright to permit a naked-eye observer to see the star or planet right up until the actual moment of hiding. If, however, the body being occulted is Venus or Jupiter, or especially if the body's disappearance is behind the unlit side of a crescent moon, then the sight for unaided vision may be spectacular.

A small telescope is necessary for most of the wonders of these closest possible conjunctions—occultations—but most conjunctions are best viewed with the naked eye. The planets, stars, and moon in various combinations of conjunctions are endless blendings of beauty. Even a pairing of planets which seems similar to one you have observed previously is likely to be a wonderfully different experience because the environment of weather and world around you, as well as your mood and mind, are likely to be different. I recall rising at 4 A.M. for a fine conjunction, to discover that the sky was overcast—except for a very small "window" of extremely clear sky in the east, through which shined the lustrous lemon crescent of moon with Venus and the bright star Regulus right beside it. And then, as I stood there watching their bright glistening in the persisting sky-window, the reverent silence of the woods and late night was touched with a quiet sound. At first I could not believe it. But then, delicately, the noise slowly increased; numerous tiny notes were sounded individually and then multiplied, blending and melting into a soft chorus. It was raining. I stood there looking at these celestial beauties shining brilliantly while all about me a gentle shower whispered on the leaves and I felt the drops. What a combination of peace and wonder! Never assume that one celestial event will be like another: each time you go out to observe, you observe not in a solitary vacancy, but in the everlastingly rich and various dance of all natural and human phenomena.

The best conjunctions of a year are so stunning that even people who ordinarily pay no attention to nature may be surprised and impressed. The years 1979–1981 were not only rare and precious for our first great visions of planets from spacecraft, they were also years of incredible grandeur for Earthbound observers of conjunctions. Some of the events which occurred were once-in-a-lifetime or possibly even first-time-in-history happenings. More glory approaches in the years before us.

Let me end this account of the planets with just a few brief glimpses of the planets from their places in the gallery of my memory of these past few years. . . .

I recall reclining moon-boat, color-tinged clouds, bright planets, streaks of meteors, Saturn spark or shard of light on one December night; two mighty planets, Jupiter and Mars, at once intense spear points and large masses of fire (like heavy eyefuls of heaven-hung flame), as they stood together opposing and conquering even the full moon, in a once-in-143-year battle of splendor on a February evening; the "touching" of a crescent moon by the long rays of Venus, followed by the first clear, flowing song of a female cardinal in the just-awakening light of a perfect May dawn; Mars and Saturn drawing closer and closer night by night, until one planet, red, was seen tucked beneath the other, gold, and they half melted into each other on a chilly, clear June night. But perhaps most of all I remember the great conjunction of Jupiter, Mars, and Regulus on the first evenings of May 1980. A glittering compact triangle of red, yellow, and blue, they hung the sky heavy with their concentration of light, shot splendor to every corner of the heavens, mingled bright beams together in glorious blending, swelled the mighty chest of Leo to the bursting brim with radiance that filled all eyes and hearts.

III.
THE STARRY UNIVERSE

CHAPTER EIGHT

Winter and Spring Stars

Have you ever stood outside on a clear, moonless night in the country? I have heard the voices of the most confirmed of city dwellers hush a little and thicken (like a sky with stars) with wonder when they told me about their great experience of that night: the night they saw the stars! Just beyond the glare of our light-wasteful cities, the stars wait still, as they have waited, most abidingly, for countless ages since man awakened, since this little Earth was first spun out of the material that had already been shining in the first generation of the great stars. That sight of the host of the stars on a dark, dark night is the *grandest* view we can have of the physical universe we live in. It strikes us with a beauty that comes naturally, not just because the stars are part of nature, but because we are, too. The vastness of the heavens, the power and size and remoteness of the stars, fills us with awe. But they should not make us feel insignificant: we are kindred to those distant lights of heaven, and though the flickering brevity of our lives seems small amidst their enduring grandeur, we are as mysterious and wondrous as they are, and at least as worthwhile.

The stars, like us, forever retain their ultimate mystery; but we are troubled by their otherness only until we become better acquainted with it. Then their strangeness is not alienating, but wondrous. Many a stargazer—amateur and professional—has written or remarked that the stars had become his friends. The comment is not an exaggeration or a frivolity. The stars that shine tonight are substantially the same that have watched over and delighted every person, happy or sad, who has ever lived. And those same stars will be there to greet you every year for the rest of your life, as they return in their season.

Fig. 36. *The main stars of the constellation Orion, mostly hot, young, and blue. This excellent photograph by Steve Albers, taken with a minimum of equipment, also shows the Great Orion Nebula and traces of other nebulosity—including faintly the elusive giant Barnard's Loop—arcing to the left of the three-star Belt.*

In these next chapters, our feet remaining firmly on Earth, we will let this planet take us on one full circuit of its orbit—a one year's tour of the heavens. The great majority of the stars and other celestial objects we will encounter can be seen with the naked eye, though certainly to see them well you need to get away from the very brightest urban illumination. If you live in a city, take a short trip—perhaps just a few miles' ride—just once a month or even once a season. If you do, and spend only an hour or two on each excursion, you can acquire a basic familiarity with almost every constellation, star, and other object mentioned here—many dozens of them—all in a year or less. But in this tour you will learn of more than just what the eye can see; you will also learn many of the facts about the stars that have been discovered by astronomers, and many of the rich tales woven about them by poets. Worlds of double suns, monsters and heroes, intricate clouds of glowing gas in which stars are born, gold and blue and "champagne shot with roses," wheels of stars by the billions which are so distant that entire wheels look like fireflies floating around the imagined tresses of a constellation queen who actually walked the Earth in ancient times: these and many dozens of other marvels we will encounter in our year's tour of the stars and our trip to the galaxies.

This tour will not include most of those wonders visible in the extreme southern region of the heavens, accessible only to people who live in the tropics or Southern Hemisphere. Yet there will be plenty. These pages are crowded with all manner of flashes of spinning suns, and the timeless images of mythic story, but they cannot possibly be as crowded with fact, dream, and glory as one small section of a night sky in the country—much less the whole arching vault of the heavens. Each night and each year you will find new friends in the starry sky, and learn more about old friends, but these chapters are just a beginning. And we begin, simply, at the start of a new year.

January is the coldest month in the northern world, but it is also the best for stars. In one relatively small area of the January sky shine four of the six (and six of the nine) brightest stars visible from midnorthern latitudes. Those stars are the most conspicuous in a group of constellations which sparkle with chilling splendor on these long, frigid winter nights. In the midst of the winter host, holding the position of honor in the very center, stands the most splendid constellation of any season: Orion the Hunter.

The stars of Orion really do suggest a vast human figure who is upright as he reaches his highest point in the south—which happens in midevening in January. The constellation contains more bright stars than any other, and they could hardly be arranged in a more striking pattern. Two stars mark the shoulders and two the knees (or knee and upraised foot) of the Hunter; between them is a trio of stars—perhaps the most eyecatching figure in the sky. These three stars, equally spaced and almost equally bright, form Orion's Belt. The Belt is located right on the *celestial equator*, meaning that it passes di-

Rastaban
DRACO
North Ecliptic Pole
Alkaid or Benetnash
Cor Caroli
Whirlpool
M51
CANES VENATICI
Chara
Alcor
Mizar
Alioth
Megrez
Phecda
Big Dipper
URSA MAJOR
Thuban
Edasich
Pherkad
Kokab
Little Dipper
URSA MINOR
North Celestial Pole
Yildun
Polaris
Merak
Dubhe
The Pointers
Owl Nebula
M97
Three Leaps of the Gazelle
Spiral Galaxy
M81
Zosma
Chort
LEO
LEO MINOR
Algieba
Regulus
Ecliptic
Road of the Sun
SEXTANS
HYDRA
Alfard
CAMELOPARDALIS
LYNX
CANCER
Praesepe or Beehive
Castor
Pollux
GEMINI
Celestial Equator
Menkalinan
Capella
AURIGA
Mebsuta
Wasat
Mekbuda
Almeisan
CANIS MINOR
Procyon
Gomeisa
M35
Galactic Anticenter
Elnath
Crab Nebula
PYXIS
Cone Nebula
Rosette Nebula
MONOCEROS
Betelgeuse
ORION
Bellatrix
Mintaka
Alnilam
Alnitak
Horsehead Nebula
Great Nebula
M42
Saiph
Rigel
PUPPIS
Naos
CANIS MAJOR
Sirius
Murzim
Aludra
Wezen
Adhara
Arneb
LEPUS
Nihal
Furud
Solar Antapex
COLUMBA
Phakd
South Celestial Pole

Fig. 37. *The starry sky for around 9* P.M. *(standard time) in January. Derived from a master chart by Guy Ottewell. © July 1975 by Guy Ottewell.*

rectly overhead at the earth's equator and can be seen from everywhere in the world.

Two brilliant stars in Orion easily outshine those in the perfect line of the Belt. The brighter is Rigel, located in the knee or uplifted foot to the west. Like most of Orion's stars, it is blue, indicating that it is very hot, young, and luminous. Rigel does not rival the star which appears brightest to us, Sirius, but that is only because Rigel is probably about 100 times farther away. Rigel belongs to the class of stars called *blue giants* and is in reality one of the most luminous stars known: if it were one of our nearest neighbors—like Sirius—it would shine with the radiance of a half-moon in our sky! Rigel has a true brightness about 57,000 times that of our own sun.

The second brightest star in Orion is Betelgeuse, located in the east shoulder of the Hunter, diagonally from Rigel. It is the brightest red star in the heavens (a beautiful contrast to blue Rigel) though it is sometimes surpassed in brightness by the red *planet*, Mars. Most observers have felt that Betelgeuse is slightly redder (or, actually, a deeper orange) than Mars, and measurements indicate that Betelgeuse is slightly ruddier than any other bright star in the sky. The measurements are not, however, strictly representative of what the eye sees, so it is a very interesting project for any stargazer to rate the colors of all the bright stars and planets on a numerical scale from very blue to very red. This is not difficult, but to do it well requires a lot of patience and a real devotion to these delicate, subtle tints.

The name Betelgeuse is an especially odd corruption of a medieval Arabic phrase (a good many of our proper names for stars are more or less from Arabic); the star itself is as odd. Betelgeuse is an unusual member of the class of stars called *red giants*. These are not so luminous as many blue giants, but they are even larger. Whereas Rigel may have a diameter about 50 times that of our sun, Betelgeuse is roughly 750 times as wide. If it were put in our sun's place, it might fill the solar system out to the orbit of Jupiter. Betelgeuse is possibly the largest star directly visible to the naked eye, but like all red giants its material is spread so thin in its outer regions that these areas are nearly a vacuum. Betelgeuse's very deep orange is a sign that the star is relatively cool and well advanced in its life.

Another interesting property of the star is directly visible to the naked eye: the variability of its brightness. Many thousands of *variable stars* are known and all stars—even our own sun!—may be subject to slight changes in brightness (and overall energy output) from time to time. Far fewer stars, however, vary markedly; of these, Betelgeuse is by far the brightest. It is usually somewhat dimmer than Rigel, but at least once—in the winter of 1852–1853—it was considerably brighter. On another occasion, it was only half as bright as its average, and scarcely outshined the stars of the Belt. The major variations seem to occur over periods of roughly six years (much longer than those of most variable stars), but there are many fluctuations and no one can predict its behavior with certainty. The star clearly bears close watching; compare its brightness to that of other stars in this region of the heavens.

Some stars are variable because they are periodically eclipsed by a fainter companion star, but the changes in the brightness of Betelgeuse correspond to actual pulsations—alterations of as much as 60 percent in the diameter of the star. That diameter is so great that even though Betelgeuse appears as a mere point of light in telescopes, the actual disk of the star has been revealed by special photographic techniques. These photographs show what seem to be brighter and darker areas on Betelgeuse, apparently true surface features which correspond to flares and sunspots on our own sun.

Not all the interesting celestial objects in Orion are stars. Just below the Belt there is a small north-south line of dimmer stars which represent the Sword of the Hunter, and around one of these you may just be able to glimpse a hazy area of light. The star Theta-1 Orionis interferes with the naked-eye view, but a small telescope shows an amazing sight: the star is actually a close-together group of stars (the four brightest are called the Trapezium) located within the spectacular Great Nebula of Orion. A *nebula* is a cloud of gas in space which may be illuminated (directly or indirectly) by stars within or around it. None is so bright as the Great Nebula (it would be prominent to the naked eye if Theta-1 and nearby Theta-2 did not hinder).

Fig. 38. *The Great Nebula in Orion, M42. Photographed with the 40-inch Ritchey-Chrétien reflector of the U.S. Naval Observatory. (See inside back cover.)*

Good binoculars show a wreath of glowing smoke, but the view in a small telescope is about the best in the heavens: the nebula's pale-green radiance is cast in a translucent fan, with brighter wisps and streamers, across velvet black and star-sprinkled space. And yet this delicate cloud of beauty is a hotbed of blazing starbirth, for it is in nebulas like this one that new stars are being born.

All of these wonders and more walk with Orion on winter nights, the mind's eye adding to the purely visual splendor. This constellation was an inspiration to the human imagination thousands of years before modern astronomy opened up new visions with telescope and spectroscope. The pattern of Orion has suggested a hunter or warrior to many cultures, but the name and myth of Orion the Hunter comes to us from ancient Greece. In those tales, Orion was the handsomest and tallest of men—so tall he could walk on the ocean floor with his head above the waves, so strong he could carry a giant Cyclops on his shoulders like a child. Perhaps the most famous story about Orion tells how Artemis, goddess of the hunt, fell in love with him, but was tricked by her jealous brother Apollo into killing Orion. When Artemis discovered what she had unwittingly done with one far-flying arrow, she was heartbroken but she gained for Orion the right to be placed among the stars as the heavens' most glorious constellation.

Orion shines above two fairly inconspicuous constellations, Lepus the Hare and the very southerly Columba the Dove. He is followed across the woods and fields of the heavens with two fine constellations which represent his two hounds: Canis Major (the Big Dog) and Canis Minor (the Little Dog). Canis Major is a noble beast, containing many bright stars. There is one marking its head (some say heart) which, on its own, commands as much admiration as Orion himself. This is Sirius, the brightest of all stars.

Sirius suffers no rivals among the stars, not even the brilliant stars of winter. It can be outshined by several of the planets but not necessarily surpassed in beauty, for, unlike stars, planets cannot twinkle. As we have seen, twinkling is caused by atmospheric turbulence (usually well above the surface); when the image of Sirius is wind-shaken it not only dances but also shoots darts of every imaginable color from its fiery heart. Color changes are with the naked eye seldom seen very prominently in other stars (they are not bright enough), but are almost always visible in Sirius when the star is fairly low (where its light has a longer path of atmosphere to traverse). The natural color of the star when it is high on a calm night is a bluish white.

It is no surprise that Sirius has been one of the most famous and revered of celestial objects. It was known as *Sopdet* (Graeco-Egyptian *Sothis*), among many other names, to ancient Egyptians, who watched for its rising just before dawn in summer to herald the life-giving annual flooding of the Nile. This *heliacal rising*—the first appearance of Sirius in the predawn sky after it has left the evening sky of winter and spring—marked the beginning of the year in Egypt. Sirius has long been known as the Dog Star (because it is in the Big Dog) and our phrase "dog days" for hot weeks in summer stems from the old belief that Sirius, near the sun in the daytime sky at that time of year, is then adding its heat to the sun's. The name Sirius, in fact, is from a Greek word meaning "scorching."

Although it cannot be seen even in many telescopes, the much dimmer companion star of Sirius is an interesting object to know about while you watch Sirius, because it is representative of the stars called *white dwarfs*. Like other stars of the class, Sirius B (playfully known as the Pup) is amazingly small for a star—about 2 percent of the sun's diameter, just a little more than twice as wide as the Earth—and yet contains as much matter as most other stars and is therefore inconceivably dense. One cubic inch of material from the Pup would weigh about 2¼ tons on Earth (but even this unimaginable density must be exceeded inside neutron stars and in the ultimate gravitational tight squeeze of black holes).

The rest of Canis Major is bright, but Canis Minor is notable almost entirely because of one star, Procyon—Greek for "before the dog" because it was seen rising just before the Dog Star Sirius. Procyon is just a little fainter than Rigel and to me has a slight tinge of yellow (but judge this subtle tint for yourself). Sirius is the closest bright star visible in the northern world, 8 light-years distant, but Procyon is only a little farther. Procyon also has a white dwarf companion which is even smaller than Sirius B. Between Sirius and Procyon is an area that lacks bright stars, called Monoceros the Unicorn. This area is interesting to the naked eye because through it runs a particularly prominent section of the Milky Way. Many people think the Milky Way can only be seen in summer, but that is merely the time when its brightest parts are visible and the greatest number of people are outside at night. If you are far from big-city lights on a January or February evening the softly luminous band can be traced all the way from the northwest to the southeast.

We find Orion and his Dogs in the southern sky and if we return to Orion's Belt—which is the center of the winter constellations—we can extend the line of the Belt eastward and come almost directly to Sirius. But if we follow the line in the opposite direction it leads us to another great constellation which is connected to Orion in star lore. On this side, Orion, who wields club and lionskin shield, is challenged by the majestic Taurus the Bull. Only the head and horns of the Bull are usually marked by stars, and some say that is because Taurus is Zeus in bull form, just half-emerged from the sea as he leaps out to carry off the maiden Europa. The brightest star in Taurus is Aldebaran, which is the Bull's Eye. This orange star, which is usually just a little dimmer than Betelgeuse, is part of a V-shaped pattern that outlines the face of Taurus. The other stars in the V are much farther than Aldebaran and are actually traveling through space together loosely in what is called a *star cluster*. This group is known as the Hyades and covers a large apparent area because it is one of the closest star clusters known—about 130 light-years distant. Most clusters appear to the naked eye—if at all—as small and dim patches of fuzzy light which a telescope shows to be a collection of stars that individually look faint and close together because they are so far away. Only a few clusters, like the Hyades, are close

Fig. 39. *The V-shaped Hyades and the dipper-shaped Pleiades are two close and beautiful star clusters in this photograph. Saturn is the bright "star" at the edge.*

enough for us to see their stars separate and bright with the naked eye. Another one of these clusters, which is even prettier than the Hyades, is located farther west in Taurus, but in that direction we are entering into the domain of the autumn stars—so we will stop to look at that cluster at the end of our year's journey, when we will have encircled the entire heavens.

If you extend lines from the arms of the V formed by the Hyades and Aldebaran, you are tracing the horns of Taurus, each of which is tipped with a fairly bright star. Next to the dimmer, more southerly star is the Crab Nebula, a cloud which can only be seen in telescopes but which is very famous in astronomy because it is the well-observed remnant of a mighty star explosion—a *supernova*—which occurred in A.D. 1054 and which for a while made a faint star into an object as bright as Venus. The brighter, more northerly "horn star" is El Nath, which officially belongs to Taurus but also is used to help form the pattern of another constellation, Auriga the Charioteer.

Auriga is usually pictured as a man on his wagon, holding a goat and its kids. The mother is marked by the star Capella (Latin "little she-goat"), which is second only to Sirius in brightness among the winter stars, just outshining Rigel. Capella is a distinctly yellow point of light not much different in color (and therefore surface temperature) from our sun. It actually consists, however, of two almost twin stars which are both many times larger than our sun. These two, Capella A and B, form a *double star*—the generic name for any two (and sometimes, in loose usage, more than two) stars which appear very close to each other in the sky. Double stars are usually so close together that the naked eye sees them as one point of light, and they are only seen separate in telescopes. Some double stars are *optical doubles*—merely along the same line of sight and not gravitationally bound, because one star is many light-years farther away than the other. Most double stars are, however, *binary* stars: two or more stars gravitationally bound to each other and often revolving together around a common center of gravity. Capella A and B form a binary, and the Sirius and Procyon pairs are binaries. Each of these consists of stars revolving around a *barycenter*, a common center of gravity, but Theta-1 Orionis is a binary in which the members are only traveling together through space with a close gravitational bond (not circling each other). In the case of Theta-1, the four stars of the Trapezium, and at least a few others, all journey together around our galaxy with the Orion Nebula. Theta-1 can therefore be called a *multiple star*, having more than two in a system. What is the difference between a small star cluster and a multiple star? It is really only one of degree, but in practice we find that even the most widely spaced multiple-star systems are typically more compact than star clusters.

Leaving Capella and the triangle of fainter stars which mark the she-goat's three young (and are called the Kids), we pass back east across the horns of Taurus and find ourselves just above Orion, at the most northerly point of the ecliptic. This is the spot where the full moon rides at the December solstice and where the sun blazes at the June solstice—and it is the boundary line between Taurus and the next zodiac constellation, Gemini the Twins. The Twins in Greek mythology were Castor and Polydeuces (Latin *Pollux*), brothers to most beautiful Helen, and according to one version Castor and Clytemnestra were hatched from one egg, Pollux and Helen from another laid by their human mother Leda, who had been ravished by Zeus in swan form. The two brothers were skilled in all areas of athletic and martial endeavor, and when the mortal Castor was slain the immortal Polydeuces refused to come to heaven on Mount Olympus without his beloved brother. Zeus was moved and, in one version of the story, he allowed the Twins to be together forevermore, spending half their time on earth and half in heaven. The constellation of Gemini is appropriately dominated by two stars of almost equal brightness, just a few degrees apart, which have been named after the mythical brothers.

These two stars, Castor and Pollux, are not much farther apart than the length of Orion's short Belt. Pollux is the more southerly of the pair, closer to the ecliptic, and is somewhat the brighter. It is slightly orange and has always been considered one of the stars of *first magnitude*—that is, the category of greatest brightness. Castor, on the other hand, falls just short and is traditionally rated as *second-magnitude* among those classes of brightness which proceed on down to *sixth-magni-*

tude—the faintest stars which can be seen with the naked eye. This ancient method of dividing stars into six rough categories of magnitude has been improved in modern times by a more precise magnitude scale and photoelectric measurements. In modern astronomy, a difference of exactly one magnitude is 2.512 times (so that 5 magnitudes are a difference of 100 times) and decimals are used so that Pollux is said to be not quite of magnitude 1.0, but rather only of 1.16 (remember, the higher the number, the *fainter* the star). Castor is 1.59, somewhat brighter than Polaris, the North Star, which is only 1.99. Stars that are brighter than Pollux, like Rigel and Capella, receive magnitude numbers that approach 0, and in the cases of the very brightest stars and planets, negative numbers are used—Sirius is −1.42 and Venus can be as bright as −4.6. (For a representative list of magnitudes for many objects, including the sun, moon, planets, comets, fireballs, and galaxies, see Appendix Four.)

Though Castor is dimmer than Pollux it is perhaps the more interesting star, because it is actually a system of at least six stars, two of which make a pretty pair in a small telescope. Gemini, which contains several other bright stars and a naked-eye star cluster, is the last major constellation of winter. As the first winds of spring begin to blow, and nights become shorter than days, the great host of winter constellations marches on into the west with its leaders, seven of the fifteen traditional first-magnitude stars. Bringing up the vanguard of spring in the south sky are three celestial beasts, two of them members of the zodiac following Gemini. These two are dim Cancer the Crab and mighty Leo the Lion, and under them stretches the foremost section of long, long Hydra the Sea Serpent.

Cancer the Crab is one of the faintest constellations of the zodiac but at its center lies one of the finest star clusters in the heavens, the remarkable object nicknamed the Beehive. This cluster is easily visible to the naked eye, though it is best observed with binoculars or very low magnification in a telescope, where it earns its nickname by appearing as a swarm of dozens of little stars, with many of the brighter ones bunched and paired and of similar brightnesses. The overall magnitude is about 3.7, with stars spread over an area roughly $1\frac{1}{2}°$ in diameter (about 3 moon diameters). The Romans called this cluster *Praesepe*, which means "manger," for the two naked-eye stars adjacent to it were the Aselli, young asses coming to the manger. Although the Beehive has been known throughout history, it received its more technical name, M44, in the eighteenth century, when it became the 44th entry in the list of clusters, nebulas, and galaxies compiled by French comet hunter Charles Messier. Messier's list was originally intended as a compendium of celestial objects which might be mistaken for the fuzzy head of a new comet, but to modern amateur astronomers the compilation of *Messier objects* or *M-objects* is a handy list of many of the finest sights visible in a small telescope. Moreover, the naked-eye view of Messier object 44 should not be neglected: in dark skies the brighter members of M44 are on the verge of detectability despite the distracting proximity of the other members. At the very least, one is sometimes able to see that this large, hazy patch of luminosity does not glow with equal brightness across its entire area—a beautiful effect. The Beehive is also interesting because its center is only 1.3° away from the ecliptic—and is thus a frequent target for planets and asteroids to pass near or even through.

The next constellation of early spring is as prominent as Cancer is inconspicuous. Leo the Lion is one of the few constellations whose pattern really does suggest the thing it is supposed to represent. Leo is composed of a fine assortment of bright stars, but the brightest of them is the particularly interesting star Regulus, which marks the mighty heart of the Lion.

The name Regulus is Latin for "little king" and was given to this star by the great Nicolaus Copernicus. Regulus has a magnitude of 1.36 and is the least bright of the stars considered first-magnitude in the traditional scheme. This blue star is even closer to the ecliptic than the Beehive (just 0.46° off) and no bright star is so often involved in interesting conjunctions with the moon and planets. Regulus also forms the base or handle-end of an asterism called the Sickle (an *asterism* is any pattern of stars which is not an official constellation; it is usually part of one, as with the Sickle and Orion's Belt). The Sickle or backward question mark of stars which curves up from Regulus

CASSIOPEIA
Caph
Shedir
CEPHEUS
Alderamin
Alfirk
Errai
Deneb
LYRA
Vega
Ring Nebula
Solar Apex
North Celestial Pole in 13,000 BC & AD
URSA MINOR
Polaris
Little Dipper
Yildun
North Celestial Pole
Kokab
Pherkad
Eltanin
Rastaban
DRACO
North Ecliptic Pole
Thuban
Edasich
HERCULES
The Keystone
M13
Rasalgethi
Rasalhague
CORONA BOREALIS
Gemma
SERPENS (CAPUT)
Unukalhai
Celestial Equator
BOÖTES
Arcturus
Alkaid or Benetnash
Mizar
Alcor
Big Dipper
Alioth
Megrez
Phecda
Merak
Dubhe
The Pointers
Owl Nebula M97
Spiral Galaxy M81
Whirlpool Galaxy M51
Chara
Cor Caroli
CANES VENATICI
M3
North Galactic Pole
Blackeye Galaxy M64
COMA BERENICES
Three Leaps
Zosma
Chort
Denebola
Realm of Galaxies
Pinwheel Galaxy M99
Vindemiatrix
VIRGO
Porrima
Spica
LIBRA
Zubeneshamali
Zubenelgenubi
Ecliptic
Road of the Sun
HYDRA
Sombrero Galaxy M104
CRATER
CORVUS
Menkent
CENTAURUS

Fig. 40. *The starry sky for around 9 P.M. (standard time) in April. Derived from a master chart by Guy Ottewell. © July 1975 by Guy Ottewell.*

is supposed to represent the powerful head, neck, shoulders, and mane of the Lion. The figure of Leo is completed by a nearly right-angle triangle of stars which form his hindquarters. The two stars in Leo which are almost as bright as Regulus are Algieba, a close-together double star in the Sickle, and Denebola, the star which marks the eastern end of the Lion.

Leo is an appropriate stellar rendering of any great lion of legend, but he is most often associated with the fearsome Nemean Lion which Hercules had to vanquish as one of his great Twelve Labors. It is said this Lion fell from the moon as a

Fig. 41. *A totally eclipsed moon is seen among the stars of Leo the Lion.*

meteor to the isthmus of Corinth, where it ravaged the countryside. The Nemean Lion was invulnerable to all weapons, but Hercules grabbed the giant beast and strangled it with his mighty clasp.

Another monster which it was Hercules' task to defeat was the one represented by the constellation below Leo, Hydra the Sea Serpent. To call Hydra a "sea serpent" is not strictly accurate, because in Greek mythology hydras were a very special kind of monster, giant snakes with numerous heads. The one which Hercules challenged was the most terrifying, the Lernean Hydra, whose venom was the deadliest in the world (there was no antidote). This monster had nine heads and chopping off one only resulted in the immediate regrowth of two new ones in its place. Worst of all, one of the nine heads was immortal! How could such a creature be overcome? Hercules enlisted the aid of his nephew Iolaus, who prevented the regrowth by burning the stump of each neck as soon as Hercules had chopped it off. And the immortal head? Hercules rolled a boulder which no one else could even budge right on top of the immortal head, trapping it forever.

As a constellation, Hydra is the longest of all, measured from east to west. When the celestial Hydra's one head is beginning to lower into the southwest, its tail is still rising on the southeast horizon. Hydra's stars are mostly faint, but the head itself (presumably the immortal one) is rather prominent and beautiful, and actually does resemble what it is supposed to represent. The brightest star in Hydra is his heart, a second-magnitude star which is orange or slightly reddish. Its name is Alphard, which means "the lonely one" and is appropriate because the area surrounding this part of Hydra, especially to the south and west, is almost devoid of prominent stars.

So far our journey among the constellations has taken us through the southern sky, but in midspring there is good reason to look north. High in the northern sky, just across the zenith from Leo, hangs the most famous of all star patterns, the Big Dipper. Orion is considerably brighter but the Hunter is visible in the evening only during the winter and early spring months whereas observers at fairly high northern latitudes (where most of the world's population lives) can see the Big Dipper every clear night of the year. The key to this seven-star pattern's eternal visibility is its location near the north pole of the sky. The *north celestial pole* is that spot among the stars which the north end of the Earth's axis points to. Since the Earth rotates about the fixed axis, any star at or very near the north celestial pole—such as the famous Polaris—will also appear fixed, motionless (or nearly so), while the planet turns onward and the night progresses. All the stars slightly farther away from the north celestial pole—such as the Big Dipper—will appear to move in medium-sized circles about that point, wheeling about forever above the horizon, while the stars very far from the pole will describe circles so big that they are cut off by the east and west horizons (and thus rise and set). The number of constellations which stay above the horizon all the time—and are called *circumpolar*—varies according to the latitude of the observer, but all of the Big Dipper is within about 40° of the north celestial pole and so constantly above the horizon for anyone who is at 40°N latitude or higher on the Earth. The north celestial pole and the North Star, Polaris, are always found exactly as many degrees above the north horizon as there are degrees in the observer's north latitude. Therefore, at 40°N, about the latitude of New York City, Polaris is 40° above the north horizon, and even when lowest, the Big Dipper is still scraping along the edge of land and sky.

Spring is the time when the Big Dipper is highest in mid-evening, and it is actually possible to tell time by the pattern's position (in any season of the year). The Big Dipper also, of course, functions as a device for finding Polaris and thus true north. Simply take the outer two stars (the Pointers) of the bowl (those opposite the handle) and extend a line through them up from the bowl (up, that is, from what would be pictured as the bottom of the bowl). The first bright star met along this line is Polaris.

The Big Dipper, despite its fame and prominence, is not an official constellation—it is an asterism, the most conspicuous part of the much larger Ursa Major, the Great Bear. In the figure of the Great Bear, the Big Dipper's bowl becomes its hindquarters, the handle its tail. The only problem is that bears do not have long tails! It is interesting how many different cultures have imagined (in some cases, no doubt, independently) that the stars here represent this animal. Many American Indian tribes did, but they were certainly not foolish enough to give their celestial bear a long tail. To some of them,

Fig. 42. *The aurora borealis and the Big Dipper as they appeared from Glacier National Park, Montana, on the night of August 11–12, 1978. Photographed by Dennis Milon with a 15-second exposure.*

Fig. 43. *The motion of stars near the north celestial pole is demonstrated in this one-hour exposure taken by George East on August 8, 1975, at Mashpee, Massachusetts, using Fujichrome 100 and a 55-mm lens at f/4.*

the three-star Big Dipper handle (or bear tail) was three hunters who were pursuing a bear (made from the stars of the bowl). In one particularly beautiful version, the stars of the handle were birds who were hunting the bear, accompanied by other bird-stars in this area of the sky. The innermost star of the handle was Robin, who at last succeeded in wounding the bear in autumn, the season when the Big Dipper is lowest (in the early evening). Robin's breast was stained with the blood, which was so profuse it even ran down to dye the autumn woods below.

Of the seven main stars of the Big Dipper, six are of second magnitude (the other is 3.3). Perhaps the most interesting is Mizar, the star at the bend of the handle. The point of light we see is itself a close-together double star (separable with a small telescope), but this pair has a third companion which is far enough from them to be distinguished by average eyesight on a clear night. The little companion star is Alcor, which the Norsemen said was the frozen toe of Orwandil (our Orion) snapped off and cast here by Thor. Mizar and Alcor together have been popularly known as Horse and Rider.

The Pointers, Merak and Dubhe, lead to Polaris, which is the tip of the Little Dipper's handle or, more properly, the tip of the unnaturally long tail of Ursa Minor, the Little Bear. This constellation may have been formed about 2,600 years ago (much later than most of the major constellations) by the first scientist, Thales of Miletus. Although the Little Dipper is rather inconspicuous on the whole, two of its other stars are fairly bright (one nearly equals Polaris) and are called the Guardians of the Pole. If Ursa Minor was not recognized as an independent constellation before Thales, it is no doubt because in very ancient history it did not contain the north celestial pole within its bounds. The exceptionally slow migration of the north celestial pole among the stars is caused by a constant changing in the *direction* of the earth's axial tilt called *precession*. The *amount* of tilt of the earth's axis does not change (it remains 23½°, and so we have our traditional seasons), but the collective tug of the other bodies in the solar system causes the axis to fall in a slow circle 23½° away from the line perpendicular to the plane containing the Earth and sun. One of the results of precession is the movement of the north celestial pole in just such a circle ($23\frac{1}{2}°$ in radius) among the stars. The motion is so slow it takes 25,800 years to complete one circle. But in the few thousand years of man's recorded history the north celestial pole has trekked from the constellation Draco the Dragon into Ursa Minor, past the Guardians of the Pole, and finally to its present position, near, but not exactly at, Polaris. The effects of precession and the polestars of past and future will be discussed further at our meeting with Draco in Chapter Nine.

There are several other prominent constellations in the north circumpolar region, but we will turn to them in the seasons when they are highest (above Polaris) in the early evening. Our return to the southern sky is conveniently accomplished by extending the curve of the Big Dipper's handle and following this arc until we reach a brilliant star, the brightest of spring, Arcturus.

The constellation in which Arcturus shines is Boötes the Herdsman, but the star's name was probably once applied to the entire figure, for Arcturus means "guardian of the bear"—that is, guardian of his flock or herd (which some say are all the other constellations) from the dangerous Great Bear. Assisting Boötes in his task is Canes Venatici, the Hunting Dogs, a rather inconspicuous constellation under Ursa Major's tail. The Herdsman himself is formed by a pattern of stars which looks somewhat like a kite flying high on spring winds.

Arcturus was once considered to be a little fainter than Capella and the summer star Vega, but recent measurements indicate that it shines at a negative magnitude of -0.06, while Vega and Capella are dimmer, at 0.04 and 0.06, respectively. Since even experienced variable-star observers find it virtually impossible to distinguish differences of less than 0.1 magnitude between two objects, it is easy to see why there has been uncertainty about the relative order of visual brightness for these three bright gems. Arcturus is a distant second to Sirius among stars visible from 40°N latitude, and fourth in all the heavens, after Sirius and two stars only visible from more southerly latitudes, Canopus and Alpha Centauri (strangely, Canopus is almost due south of Sirius, and Alpha Centauri is almost due south of Arcturus in the sky).

Arcturus is usually said to be orange but it is bright enough to make some sensitive observers complain that the term is too rough. One observer I know says that the color of this star is "champagne shot with roses." Arcturus is also distinctive in ways other than its brightness and color. Whereas most stars we see are traveling more or less like the sun in the central plane of the galaxy, Arcturus is an old gypsy star which happens to be piercing south through this plane on its tilted orbit around the galaxy. For long ages, Arcturus journeys far above or below the plane, so we are fortunate indeed to live in a time when it is passing close enough to be one of our nearest bright neighbors. The distances of interstellar space are so great that most stars do not change their positions in relation to other stars—and hence do not alter the shapes of the constellations—more than the tiniest amount in thousands of years. But Arcturus has moved noticeably even within historic times, and our earliest human ancestors did not know this star because it was then too distant to be seen with the unaided eye. In the future, Arcturus will race far south in our skies and in 500,000 years—an extremely short time in the life of the stars—it will have receded again beyond the seeing of the naked eye on Earth.

When we gaze in the direction of Arcturus we are indeed looking upwards out of our galaxy, perpendicular to the central plane. As a result, the spring constellations do not have many bright stars. The region of Arcturus is especially sparse, so that the star appears even more prominent by virtue of hanging in lone splendor. Since there are far fewer stars and fewer clouds of interstellar gas and dust here above our galaxy's plane, it is possible for us to look out and see other galaxies floating in space, unimaginably far away. In the region between Arcturus and Leo is the constellation of Coma Berenices, Berenice's Hair. The main pattern is formed by a delightfully irregular little scattering of stars which is one of the closest and brightest star clusters. All around the star cluster, we find part of a far vaster cluster (vaster in both the area it covers in the sky and the actual volume of space it occupies): a cluster of distant galaxies (we shall visit them in the last chapter). The legend of Berenice's Hair is based on historical personages and may be largely factual. The story says that Berenice, the queen of Ptolemy III of Egypt (who reigned 246–221 B.C.), vowed to cut off her beautiful "amber tresses" and offer them to the gods if only they would let her husband return safely from war. She made good on her promise and the sacred locks were placed in the temple—but soon after, they mysteriously disappeared. Happily, the court astronomer Conon was able to solve the mystery: Berenice's hair had been accepted by the gods and placed to shine as a new constellation in the heavens. There we see it on spring evenings even unto today.

South of Coma, the wonderland of galaxies (visible only in telescopes) continues and reaches its greatest richness in the western region of a very large constellation, Virgo the Virgin. This maiden is usually identified as Ceres (the Greek Demeter), the goddess of harvest and fruitfulness, of all living things that grow in the earth. When Virgo is last seen in the evening sky, setting just after the sun, autumn has come and it is harvest time. Virgo-Ceres is the figure described by Keats in his great ode "To Autumn": the constellation is usually pictured as a woman, sometimes with wings, who is reclining and holding an ear of wheat in her left hand, at her hip. The ear of wheat is a symbol of plenitude and fulfillment, and is marked by Virgo's *lucida* (brightest star), Spica (a name which literally means "ear of wheat"). At magnitude 1.00, white or blue-white Spica is a little brigher than Regulus and is close enough to the ecliptic to be involved in many conjunctions (Virgo, of course, is a member of the zodiac).

If we continue south from Spica we have almost completed a vast, sky-filling loop of the spring constellations which started in Cancer, Leo, and Hydra, went round through the northerly Bears, and descended through Arcturus (in Boötes) to Virgo: for here, below the harvest maiden, is at last the end of Hydra's prodigious length, the tail which we left to be continued. It is a faint part of the great monster, but tucked between it and Virgo is a pair of notable constellations: Corvus the Crow and Crater the Cup. These two, which are often pictured resting on the very coils of Hydra, have often been associated with each other in legend. Perhaps the most famous story about them is an Aesop fable in which a thirsty Crow sought to quench his thirst from a goblet of water. The level of water was too low for the bird to reach, but crows are very clever and Corvus was no exception: he got pebbles and continued to drop them in the goblet (Crater) until the water level was high enough for the wise bird to drink his fill. As constellations go, Crater is quite faint, but Corvus is formed by a fairly small quadrangle of stars which are magnitude 3.0 and brighter and are a conspicuous sight in this largely dim region.

Our journey through the stars of winter and spring comes to a fitting close—to the very gates of summer—with a constellation that brings together myth and science particularly well. Flying back from the Crow as the eye flies, we pass Arcturus, now shining high in the southwest as night falls after a long June day. Just to the east of Boötes, and very high in the south at this hour, is Corona Borealis, the Northern Crown.

This Corona certainly is one of the crowning glories among constellations. None of its stars are extremely bright, but those that are not faint are arranged in beautiful adjacency, in a sort of semicircle which does resemble a fillet or crown. The ancient Greeks said this celestial ornament belonged to the maiden Ariadne, who helped Theseus escape from the Labyrinth but was abandoned by him—only to be rescued by the god Dionysus. The god was faithful to her, and after her death he tossed the crown he had given her all the way up to the heavens where it shines forever in her honor.

The brightest star in Ariadne's crown is Gemma, also called Alphecca, a pretty one almost as bright as Polaris. There are several notable variable stars in the constellation, but one of them must be considered among the heavens' strangest and potentially most exciting objects. It is T Coronae Borealis, a star which is normally too faint to see without a telescope but which in 1866 flared up to 2,500 times its normal brightness, outshining Gemma and earning a nickname: the Blaze Star. After its startling outburst the star quickly faded back to its original brightness—so quickly that some astronomers thought that T could not be a *nova*. A nova is an exploding star in which about 10 percent of the star's mass is blown off in a brilliant outburst; a *supernova*—such as formed the Crab Nebula in Taurus—is a far brighter and more catastrophic event in which perhaps 90 percent of the star's mass is lost. The Blaze Star might be a bright nova, but the rapid decline after its great flaring was unlike that of any previous nova. Nevertheless, exploding stars of any kind were believed to be one-time affairs and it seemed that T Coronae could surely be relegated to the status of past wonder. The nature of T Coronae was still being debated in 1946, when suddenly there was some new data: the Blaze Star had exploded again!

T Coronae Borealis is now listed as the brightest member of a small class called *recurrent novas*, though no one yet knows if the title is really appropriate, for the mechanism involved may be quite different from that in true novas. Astronomers do know that T Coronae at maximum shines with a luminosity that greatly exceeds even such mighty blue giants as Rigel. The rise and fall of the Blaze Star are so swift that in 1946 the peak was apparently missed, the star already having dimmed back to magnitude 3.2. It is even possible that several explosions of T Coronae—perhaps even in recent years—have been missed altogether. Therefore, it is an excellent idea for any stargazer to learn the beautiful pattern of Corona Borealis and admire it each spring and summer night, checking to see if there is any new gem in the semicircle. The Blaze Star could flare forth again at any time.

Few constellations have inspired legends in more cultures than Corona Borealis. The Australian aborigines, not surprisingly, identify it as a boomerang. But an especially rich and peculiar legend about Corona is told by the Shawnee Indians. They saw these stars as an incomplete circle of dancers, eleven beautiful heavenly maidens who when they had been twelve in number had sometimes come down to earth in a great silver basket to dance upon the grass. One day, long ago, the twelve were seen by the young warrior Algon (White Hawk), who hid and watched them long in delight. Each time he returned to spy on them, he fell deeper in love with the youngest and fairest of the sisters, until finally he magically transformed himself into a field mouse and crept up into the circle, where suddenly he changed back and seized the young maiden. The other sisters fled, grief stricken, but the young starmaid, at first sad, soon learned to love Algon. They were very happy for a few years, and had a son together, but at last the woman was so filled with longing to see her sisters and her fair home again that she secretly resolved to leave. Therefore, while Algon was away on a hunting trip, she built a magic basket and returned to the field where she and her sisters had once danced. Algon came back just in time to see the basket flying into the distance with his wife and son inside. But the story has an altogether happy ending: the starmaid missed her husband dearly, and so at last her father granted Algon the right to come to heaven-world, which he did, bringing tokens of the animals up to the starfolk as gifts. Algon and his wife and child now have the freedom of both worlds, between which they can be seen journeying as white hawks or falcons; when Algon is in the heavens he is Arcturus, and the eleven sisters, though they are not forever separated from his wife, must dance without her, so we see Corona Borealis as an incomplete circle.

This wonderful tale leaves us with two puzzling questions. First of all, why are there eleven maidens when there are only seven stars clearly visible in the semicircle? There are several possible answers, including the possibility that twelve was a number of some mystical import to the storytellers. But a more interesting question is why (to the best of my knowledge) there is no word on where the star of the young maid now is. Could she be the Blaze Star? The Blaze Star would be youngest (last "born") and fairest (brightest) of the stars of Corona at maximum. Could this beautiful story be in part a remembrance—however altered by traditions and embellishments—of an actual observation of some past outburst of T Coronae? It seems very unlikely, but the speculation is fascinating.

As spring ends, Orion has long since disappeared and the very last of winter's brilliant host is descending in the west, even the Lion of early spring making his way towards the horizon. But on these short June nights, there is a blaze called T that is quietly preparing to explode; a queen's crown and a herdsman; a big orange gypsy star that is possibly on a kite; a goddess of the harvest with a glowing ear of wheat; crow and cup and the serpent slithering in the south; Berenice's galaxy-surrounded tresses; heavenly sisters circle-dancing.

And the brilliant summer Milky Way, with a whole new glorious and various assemblage of constellations, is climbing in the east.

CHAPTER NINE

Summer and Autumn Stars

In the last chapter, we halted at the brink of summer, near the zenith or crown of the sky, with Corona Borealis. The Earth has swung around since January to the other side of its orbit, so that constellations like Orion, which in January appeared opposite the sun (high in the midnight sky), are now in the same direction as the sun (high in the midday sky), and are thus lost in the solar glare. If the sun did not hide the stars in the day sky, we would now see it among the stars of Gemini and Taurus, at the most northerly part of the ecliptic: the sun is at the June solstice and thus it is the first day of summer in the Northern Hemisphere. At this time, Orion is just below the sun, rising with it at dawn and setting with it at sundown; each day hereafter Orion will rise four minutes earlier, slowly (as the summer progresses) becoming visible in the east before sunup. If you want to see stars of winter in the late summer, you need only go out to observe very late, in the predawn hours. But our journey—through the year and around the heavens—is based on what we can see, conveniently, before midnight in each successive season. So in this voyage among the stars and through their timeless myths, the next stage is the constellations of summer.

The bright summer Milky Way is now lifting its great arch in the east, and already visible along its length are the noble constellations from Cygnus to Sagittarius. But arriving ahead of those richest sights of the great Heavenly River are important constellations of its western shore.

Arcturus burns brightly in Boötes with Corona Borealis just to the east of the Herdsman (in earliest legend, our Northern

ANDROMEDA
CASSIOPEIA
CAMELOPARDALIS
CEPHEUS
URSA MINOR
LACERTA
PEGASUS
CYGNUS
LYRA
VULPECULA
SAGITTA
DELPHINUS
EQUULEUS
AQUARIUS
CAPRICORNUS
AQUILA
SERPENS (CAUDA)
SCUTUM
SAGITTARIUS
CORONA AUSTRALIS
Capella
Mirfak
Almak
Mirach
Great Galaxy M31
Sirrah or Alpheratz
Double Cluster
Ruchbah
Shedir
Caph
Errai
Polaris
North Celestial Pole
Yildun
Little Dipper
Kokab
Pherkad
M81 Spiral Galaxy
Great Square of Pegasus
Scheat
Markab
Alfirk
Alderamin
North Ecliptic Pole
North America Nebula
Deneb
Eltanin
Rastaban
North Celestial Pole in 13,000 BC & AD
Vega
Ring Nebula
Solar Apex
The Keystone
Gienah
Veil Nebula
Enif
Sadalmelik
The Urn
Sadalsuud
Dumbbell Nebula
Albireo
Altair
Saturn Nebula
Albali
Celestial Equator
Rasalhague
Cebalrai
Alya
Giedi
Dabih
Ecliptic or Road of the Sun
Eagle Nebula M16
Horseshoe Nebula M17
M20 Trifid Nebula
M8 Lagoon Nebula
Galactic Center
Shaula
Nunki

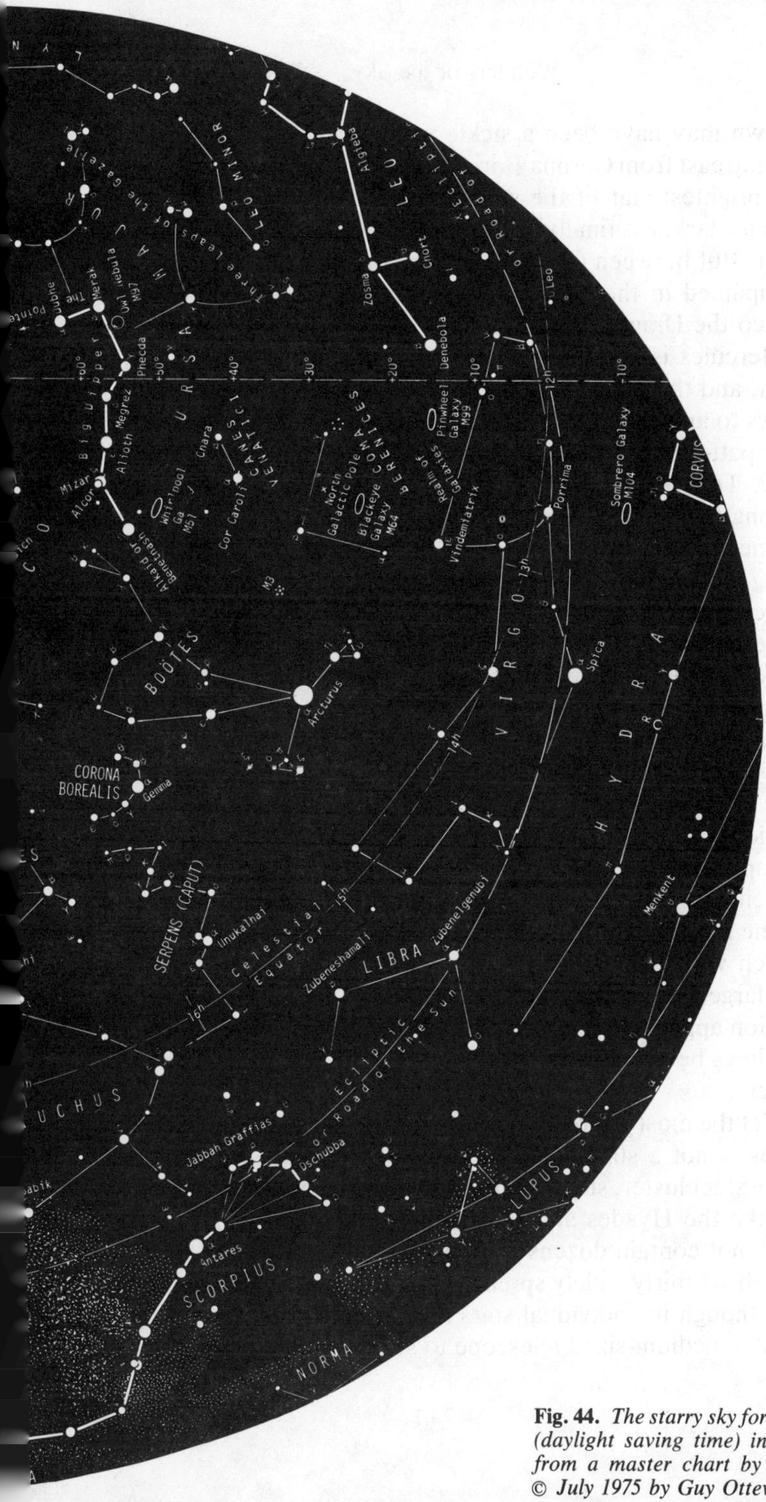

Fig. 44. *The starry sky for around 10* P.M. *(daylight saving time) in July. Derived from a master chart by Guy Ottewell. © July 1975 by Guy Ottewell.*

Crown may have been a sickle in the hand of Boötes). Continuing east from Corona Borealis eventually brings us to Vega, the brightest star of the summer, already shining high in the east as darkness finally falls at the end of this day at summer's start. But between Corona Borealis and Vega is Hercules, accompanied in the northern sky by the head of long, curling Draco the Dragon.

Hercules is certainly the most famous of legendary strongmen, and the many tales about him go back several thousand years to ancient Greece. But long before even those times, the star pattern here was a mysterious figure called the Kneeling One. That is still the posture in which we imagine Hercules among the stars, holding his club and wearing the skin of the Nemean Lion. In modern times, however, Hercules has been hung upside down among the stars, by precession (about which, more in a moment).

Perhaps the most interesting star in Hercules is the one which marks his head (at the southern end of the constellation). Rasalgethi, like Betelgeuse, is a long-period variable which is semiregular in behavior. The changes seem to be based on two cycles, one of about 50 to 130 days, the other of about 6 years. The star varies from about 3.1 to 3.9 in magnitude, so its behavior is interesting to watch with the naked eye. A small telescope shows, however, that Rasalgethi is different from Betelgeuse in one very important respect: it is a double star, one of the most beautiful in the heavens. One of the pair—the one which varies—is indeed a vast red giant, once thought to be the largest of all stars. In a telescope the fifth-magnitude companion appears blue-green next to the orange of the large star, the hues being made far more vivid by their contrast with each other.

Yet the most interesting and beautiful celestial object in Hercules is not a star, but a star cluster. It is not, however, an *open* star cluster, such as the ones we visited in the last chapter. Unlike the Hyades and the Beehive, this object in Hercules does not contain dozens of (telescopically) rather bright stars which are fairly widely spread ("open"). It is a *globular* cluster and though its individual stars appear quite faint (requiring at least a medium-sized telescope to see) there are not dozens or hundreds, but instead about *one million stars* here compressed into an area of sky many times smaller than that of the clusters we have already visited.

The great globular cluster in Hercules is M13 (the 13th object on Messier's list), and it is the most glorious of its kind plainly visible from the United States (much more southerly observers can see the even better Omega Centauri and 47 Tucanae globular clusters). The naked-eye observer with dark skies can observe M13 as a dim point of light between the two western stars of the four-sided Keystone pattern which forms the lower body of Hercules. But it is fascinating to know that this point of light is actually a ball of a million bright stars about 25,000

Fig. 45. *The great globular cluster M13 in Hercules. Photographed with the 200-inch Hale telescope.*

light-years away—many times farther than any of the individual stars we see in the sky. The stars in a globular cluster really *are* closer together than stars in open clusters, but at such a great distance they appear so much more condensed that even the largest telescopes can only separate stars in the less crowded outer regions of the cluster. The center of the cluster is a blazing mass of combined radiance dusting out into hundreds and (in larger telescopes) thousands of tiny pinpricks of light that seem to extend in starry arms or tentacles, as if this were some great (but surely benign) monster. A monumental city of stars? A celestial pile of rouge-fine jewel dust afire? No comparisons are adequate, and like the Great Orion Nebula, this is one of those celestial wonders which will awe you when your naked-eye adventures in the stars have finally tempted you into acquiring (or gaining use of) a telescope.

Just to the north of Hercules, across the zenith, is the head of the long, twisted Draco the Dragon. This prominent head is composed of four stars, one of which is second-magnitude. The Dragon's body twists first east and then back west to separate the two Dippers. A certain star in Draco is only of moderate brightness and yet is one of the most famous objects of ancient history. This is Thuban, which lies between Mizar (at the bend in the Big Dipper's handle) and the Guardians of the Pole (the two bright stars in the Little Dipper's bowl). Thuban's claim to historic fame is the fact that it was the North Star in the era around 2830 B.C. As mentioned in the previous chapter, the very slow changing of the direction of the Earth's axis called precession moves the position of the north celestial pole among the constellations, making a full circle among the stars in 25,800 years. A key shaft in the pyramid at Khufu—greatest in Egypt—was aligned to Thuban, which was then closer to the celestial pole than our Polaris now is. Whereas Thuban is not a very bright star, about 10,000 years in the future our North Star will be the brilliant Vega, though Vega will never be extremely close to the pole.

Precession also changes the position of the south celestial pole (which United States observers never see), and there is currently no bright star anywhere near this position (in the altogether dim constellation Octans the Octant). It is not difficult to reason further that, for any given latitude, precession has changed the stars which pass through the zenith. Thus Hercules was *once* right side up in the northern sky for Greek (and earlier) observers who identified the constellation, but now appears inverted in the southern sky for all observers north of the tropics. For ancient viewers the zenith lay between the head of Hercules high in the north and the head of another giant in the south: Ophiuchus the Serpent-bearer.

The pattern of Ophiuchus is very large and has adjacent to it (in fact, connected to it) the Serpent, Serpens, which is often considered to be two constellations separated by Ophiuchus: Serpens Caput, the Serpent's Head, and Serpens Cauda, the Serpent's Tail. The star in the head of Ophiuchus is rather bright (second-magnitude) and is called Rasalhague, but perhaps the most interesting distinction of Ophiuchus is the fact that it is the least well known major constellation of the zodiac. The Serpent-bearer is not one of the famous twelve given in all the traditional astrological lists but the sun spends more time in Ophiuchus than it does in its renowned neighbor, Scorpius.

There is no doubt, however, which is the more interesting and spectacular to observe. Scorpius the Scorpion is one of the few brightest constellations in the sky, and its curling S of stars does look a lot like the outline of a giant scorpion, tail raised to sting, clambering onto the western bank of the Milky Way. For United States observers, Scorpius appears quite low in the south, where summer haze is thickest, and so the constellation is never seen as well as it deserves to be. The position of this monster in the summer sky is almost opposite that of Orion in the winter heavens, and there is a myth which attempts to explain this. According to the story, Orion was killed not by Artemis, mistakenly (see Chapter Eight), but by a huge scorpion which Hera sent to punish the Hunter for his boast that he could slay any beast. When Scorpius itself later died, Hera insisted that it be placed among the stars, but Zeus hung the monster in the heavens as far from Orion as possible, so that when one is rising, the other has already set in the west.

In all the figure of Scorpius there is just one important part missing: the Scorpion's claws! But these are, in a sense, really there, just as they used to be: stars which were originally considered to form the large claws are those which now compose Libra the Scales, the constellation just west of the Scorpion, in the zodiac between Scorpius and Virgo. Even in very ancient times, Libra was an alternate way of picturing the stars of the Claws, but it did not gain the ascendancy until Roman days. And so what was once the Great Sign, largest constellation of the zodiac, became divided into bright Scorpius and rather dim Libra. But the memory of the Claws lingers yet, even in the names of the brightest stars in Libra, Arabic titles which are the most ungainly of all for English speakers: Zubenelgenubi and Zubeneschamali, the Southern Claw and the Northern Claw. These stars are more easily called by their scientific names, Alpha and Beta Librae—that is, the Alpha (first letter of the Greek alphabet) and Beta (second letter) stars of Libra. Ideally the Alpha star in a constellation should be its lucida, its brightest—but this is only the case in some constellations, for in others the determinant is position or, seemingly, supposed importance (thus, in Orion, Betelgeuse is called Alpha Orionis, and Rigel is Beta, even though Rigel is almost always the brighter).

In Libra, the brighter star is Beta—Zubeneschamali—but it outshines Alpha Librae only very slightly. Strangely, however, the second-magnitude Beta seems to have appeared much brighter to some of the ancient Greeks. Eratosthenes called it the brightest star in the entire combined Scorpion and Claws, which would make it easily first-magnitude and maybe four or five times brighter than it is at present. Modern studies have detected no indication of variability in the star, so this matter remains an unsolved mystery. Another peculiar characteristic attributed to Beta Librae is its color: according to some observers, it is distinctly greenish, even to the naked eye—the only star reputed to be so. But the tints of stars are elusive: it would be very interesting to have a group of people study Beta Librae and separately judge its color (at least a subset of the group should be formed of people who have no preconceptions about what the star's color might be).

Our present-day Scorpius, though clawless, is, as many have said, a striking constellation. Most notable of all its stars is its lucida, which is conspicuously flanked by two stars of moderate brilliance: this is Antares, the red giant which marks the heart of Scorpius. Antares shines at an average magnitude of 0.92, so that it is usually somewhat less bright than Betelgeuse. It is probably also a little bit less of a deep orange, but the color is strong and even the star's name is a reference to this: Ares was the Greek god of war, equivalent of the Roman Mars, and hence the star's name is "anti-Ares"—that is, "the rival of Mars," a planet which of course is well known for its ruddy hue.

Antares is hundreds of times bigger than our sun and, though usually just slightly variable in brightness (and, presumably, size), it has, on rare occasions, dimmed almost a full magnitude, down to 1.8. In neither size nor brightness nor range of variability is Antares quite as impressive as Betelgeuse, but it has some marvelous distinctions of its own. Of these perhaps the most interesting is its companion, which is quite close to the much brighter main star, but can be glimpsed well in telescopes with mirror or lens as small as 4¼-inch on very calm nights. The companion, Antares B, is notable for appearing vivid green beside the orange-red flame of the primary—a spectacular sight! Unlike Beta Librae, the companion of Antares appears distinctly green to most observers—but some astronomers maintain that here, as in other cases (such as the pale green-blue of Rasalgethi B), the color is an illusion caused by contrast with the main star. And yet in lunar occultations, when Antares B emerged from behind the moon first, and was seen on its own for about 5 seconds, observers reported that it still appeared distinctly green. Whatever its true color, this sixth-magnitude companion is beautiful beside its blazing primary—but also beautiful is an Antarean contrast visible to the naked eye: the deep orange of Antares compared to the gleaming white of the Milky Way nearby.

The Milky Way washes across the tail of Scorpius, and just behind the group of bright stars which form the stinger are two beautiful open clusters, M6 and M7. The latter is about twice as wide as the moon and can be seen with the naked eye in dark skies as a bright glowing patch. And once you are looking at these big clusters your eye has already been drawn into the widest and most spectacular part of the Milky Way visible from the United States. That magnificent region has its heart—the heart of our galaxy—in the Scorpion's neighbor, Sagittarius the Archer.

The patterns and gradations of brightness in the summer Milky Way are of dreamy grandeur. They are dealt with in the next chapter, which also discusses what it means to say that here, between the tail of Scorpius and the main pattern of Sagittarius, lies the center of our galaxy. The wonders of the individual stars, nebulas, and star clusters in the Sagittarius area are a more than sufficient subject for our present tour: no other region in the sky possesses quite so rich an appearance for naked eye or telescope.

The constellation pattern of Sagittarius is itself, like Scorpius, bright and memorable. Most of the major stars form the Teapot, which has a steam of nebulas rising from its westward-pointing spout. Also formed by some of the prominent stars of Sagittarius is an asterism called the Milk Dipper—placed indeed in the milkiest region of the heavens. But the stars of Sagittarius can also be arranged to outline the figure of an Archer drawing back his bow, with arrow supposedly aimed at Scorpius, perhaps at Antares itself. This Archer is a centaur—half man, half horse—and is often said to represent the wisest and most peace-loving of that usually boisterous and lecherous species: Chiron, teacher of several of the great Greek heroes (Chiron is, however, sometimes identified with a more southerly constellation, Centaurus). Not far below Sagittarius you may notice a rough circle of stars, quite near the horizon for United States observers. This is Corona Australis, the Southern Crown, which is not so bright as its northern counterpart, and yet was well known to ancient observers, for whom it was (because of precession) higher in the sky.

The star clusters, star clouds, and nebulas of Sagittarius seem almost as numerous as the individual stars there, and quite a few of them can be glimpsed with the naked eye or binoculars. Most conspicuous, besides the large star clouds of Sagittarius and nearby Scutum, is the smaller (2° × 1°) patch of M24. A small telescope shows more individual stars here than in perhaps any other spot in the heavens—on a clear night the field of view is simply packed with a wealth of star upon star that is almost unbelievable. But the naked-eye view of M24 in its breathtaking setting is maybe as good: the little star cloud appears as an intense knot of luminosity among the lights of the individual stars and the glows of other star clouds and brilliant areas of the Milky Way. M24 is located well above the western side of the Teapot, and another conspicuous patch of brightness lies almost directly north of the tip of the Teapot's spout (marked by Gamma Sagittarii). This somewhat smaller glow—about ½° across—is not, however, just the combined radiance of an especially rich area of Milky Way stars. It is the Lagoon Nebula—M8—second only to the Great Orion Nebula in brightness.

Like the Orion Nebula, this cloud of gas is associated with a group of stars which help illuminate it: this, too, is a place where stars are being born. Another of these *diffuse nebulas* in Sagittarius is M20, the Trifid Nebula, just 1½° north of M8, but much fainter. The nebula M17, about 2° north of M24, is quite a bit brighter than the Trifid, but still very difficult for naked-eye observation. M17 owes its several nicknames to its form, which is irregular like all diffuse nebulas, but which is highly suggestive: thus it has been called the Swan, Checkmark, Horseshoe, and, most commonly, the Omega Nebula (from a supposed resemblance to the Greek capital letter omega, Ω). One additional celestial object in Sagittarius deserves special mention here, for it can be spotted with the naked eye in dark skies. It is M22, found just northeast of the Teapot's top (Lambda Sagittarii). This little glow is a mighty globular cluster which rivals M13 in Hercules.

If we follow the mainstream of the Milky Way north from Sagittarius, we come first to a small and inconspicuous constellation, Scutum the Shield, which is justly famous for its great star cluster, M11, and for the Scutum Star Cloud—which is anything but inconspicuous! M11 is fainter and smaller than the close open clusters which we visited in the last chapter, but contains many more stars, so that in a small telescope it is one of the most beautiful of all objects. The stars of the cluster are arranged in a roughly triangular shape, with a bright star at the apex, and have suggested the form of a wild duck in flight to some observers. But such views are possible only in a telescope. Although M11 is prominent in binoculars, it is rather difficult to observe with the naked eye—mostly because its background is so bright! For that background is the Scutum Star Cloud, one of the brightest parts of the Milky Way.

Continuing north up the Milky Way stream we find the great band of soft radiance split by a dark tongue—the Great Rift, to be discussed in the next chapter. But when we look highest on late summer evenings we see the Milky Way cutting across a wide, rich area which is marked by three first-magnitude stars of three major constellations. The three stars—Vega, Deneb, and Altair—form the very large and prominent asterism called the Summer Triangle.

Vega leads Deneb and Altair in their nightly journey to the west and also leads them in brilliance. One observer I know has called it "the Sirius of Summer"—quite appropriate considering that it is blue-white and the brightest star of its season. Though much fainter than Sirius and marginally less bright than Arcturus, Vega burns at a magnitude of 0.04, compared to 0.77 for Altair and 1.26 for Deneb. Vega also holds the distinction of being the first-magnitude star which passes most nearly overhead for observers around the heavily populated 40°N-latitude circle of the world. This star is found in the very compact but conspicuous constellation Lyra the Lyre, which is often said to be the lyre of Orpheus, greatest of all musicians. Right beside Vega is another of the constellation's interesting objects, Epsilon Lyrae. Epsilon looks slightly elongated to most people on calm, clear nights, but keen-eyed observers can sometimes see it as two separate stars of almost equal brightness. The real surprise comes with a small telescope: each of these component stars is itself split into a pair of similar gems, so that there are actually four stars, a "double double."

Lyra also contains the fascinating variable star Beta Lyrae, which is 3.4 at brightest but which once every 13 days sinks to 4.1, only to rise again to maximum and fall to a lesser minimum of 3.75. Apparently there are two suns almost in contact, with matter flowing from the heavier to the lighter, and the dimmings are caused by various eclipses of the brighter star. Compare Beta Lyrae's brightness to that of neighbor Gamma, which is 3.25.

Fig. 46. *The Ring Nebula, M57, in Lyra. Photographed with the 200-inch Hale telescope.*

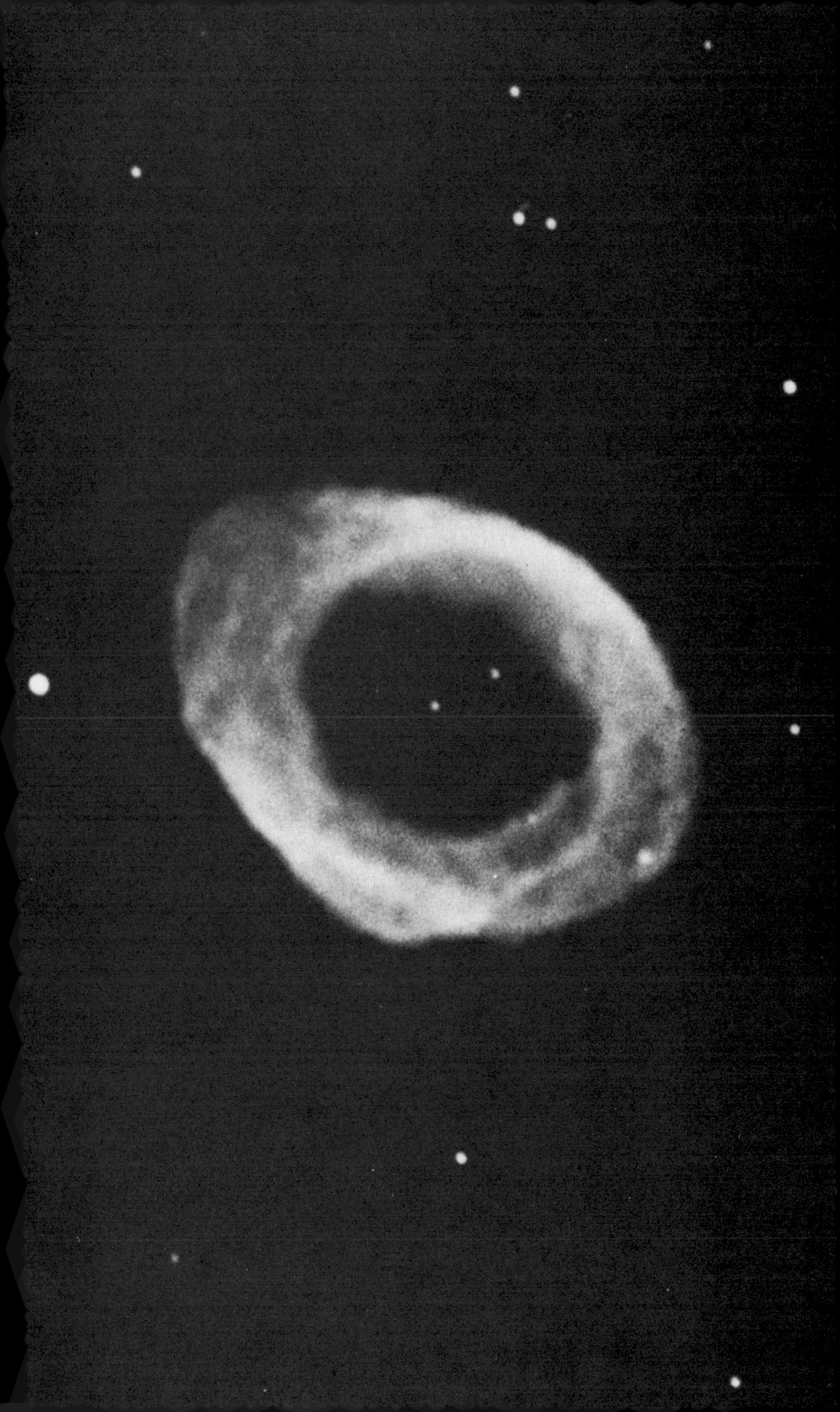

Even if you do not have use of a telescope, it is interesting to know that between Beta and Gamma Lyrae is one of the finest telescopic sights in the heavens: the Ring Nebula. Even in small telescopes, it can be seen that this object, M57, earns its name, appearing as an eerie luminous smoke ring afloat in the eternal dark. M57 is not a diffuse nebula, where stars are being born, but a *planetary nebula*, probably the ejected material from a very hot "central star" which is dying. The term "planetary nebula" originated from the fact that most of these objects are small and blue or green, closely resembling the telescopic appearance of the planets Uranus and Neptune. The Ring Nebula is much larger, the most famous and one of the brightest objects of this class.

The southern star of the Summer Triangle is Altair, a slightly yellow star which marks the head of Aquila the Eagle. Study of Altair's spectrum shows that this sun rotates faster than any bright star in the sky. Our own sun rotates also, but takes nearly a month (at its equator) to complete one full turn. Altair, which is about $1\frac{1}{2}$ times as wide in average diameter, spins one full turn every $6\frac{1}{2}$ hours! Since Altair's circumference is millions of miles, the material at its equator (where rotation is fastest) is traveling at an almost unbelievable speed. The effect must be to make Altair twice as wide at its equator as at its pole. If Altair were our sun we would see something that looked like a yellow-white egg in our sky.

Even more conspicuously than Antares, Altair is flanked closely by two stars of moderate brilliance. This trio, almost as prominent as Orion's Belt, has in some cultures played an important role in the legends of stars. One of the most famous and popular star legends of the Orient, told in many versions in different lands, is the tale of the Magpies' Bridge across the Heavenly River, the Milky Way. In the Korean account, Altair and the two adjacent stars are a Prince who has been separated by his father-in-law from his Bride, our delicate little constellation Lyra, the barrier between the two lovers being the impassable Heavenly River (which of course really does flow across almost all the space between Altair and Lyra, so that they seem to be on its "shores"). Once a year, however—on the seventh night of the seventh moon (August?)—the two lovers miraculously achieve a meeting by walking over a Milky Way-spanning bridge formed by the fluttering wings and the bodies of countless magpies who gather from everywhere on this night to help the couple. If any magpies are seen around the earth at this time, the children scold them for shamefully neglecting their duty to the Prince and Bride. After just one night, the couple must return to their state of separation, the magpies scatter back to their various homes (near and far), and the lovers must wait another year before their next meeting. A charming detail of the story is the reason for the separation, which is the father-in-law's anger at the Prince having invested (as R. H. Allen says) "the paternal *sapekes* in a very promising scheme to tap the Milky Way and divert the fluid to nourish distant stars"!

Though Aquila and Lyra are full of brightness, legend, and scientific interest, the final constellation which contributes its lucida to the Summer Triangle is possibly even better in these respects. Just as swans are among the most beautiful and magnificent of birds, so is Cygnus the Swan ranked among constellations. The main stars do form a pattern like that of a large bird with outspread wings and a long neck, and also like that of a cross—hence, another (unofficial) title of these main stars is the Northern Cross (many times larger than the bright but tiny Crux, the Southern Cross, which adorns tropical and Southern Hemisphere skies). The Swan is pointed straight down the Milky Way and heading south, which is certainly appropriate for a bird in the late summer or autumn.

The brightest star of Cygnus is the one which marks the tail of the Swan (or the top of the Northern Cross) and the name, Deneb, literally means "tail." This white star does not appear so bright as Vega or Altair, but that is only because those two stars are so much closer to us. Vega is 27 light-years distant, Altair 16, but Deneb is about 1,600 light-years away—100 times as far as Altair! Of all stars visible to the naked eye, Deneb may be the most luminous in reality. One measure of a star's true brightness is *absolute magnitude* (as opposed to *apparent magnitude*, how bright a star appears to look as a result of its true brightness *and* its distance). Absolute magnitude is the magnitude a star would have if it could be moved to the standard distance of 32.59 light-years or 10 parsecs (a parsec is another large unit of measurement). The absolute magnitudes of Vega and Altair are respectively about 0.5 and 2.2, of Sirius about 1.4, and of our own sun a meager 4.8. Deneb probably has an absolute magnitude of better than −7.0! Among first-magnitude stars, only Rigel is comparably luminous.

At the other end of the Swan from superluminous Deneb, the Swan's head or eye is marked by Albireo, which in even the smallest telescopes is revealed as a splendid double star of gold and blue. Some observers feel that this is the loveliest of all double stars, but even within the bounds of Cygnus itself there is competition. One of the other double stars in Cygnus is a fifth-magnitude orange pair called 61 Cygni. It appears as just one fairly dim point of light to the naked eye but it is worth a gaze, for when you look here you are also gazing 11 light-years to a third member of this system, which is probably a planet!

The existence of this body, 61 Cygni C, has been posited on the basis of irregularities in the motion of the two stars. It is estimated to be about 8 times as massive as Jupiter, could be smaller than Jupiter, and may require about 4.8 years to circle the A star. For only a few other stars in the heavens has evidence been found to indicate the presence of planets—but that is surely because of the limitations of our instruments. Most astronomers today believe that planetary systems are quite common, so that the chances of life—even intelligent life—among the several hundred billion other stars in our galaxy seems a virtual certainty, though no one can yet very meaningfully estimate how common it might be. Could there be intelligent life in the 61 Cygni system—perhaps an Earth-sized moon circling the giant planet? Imagine the astonishingly complex and grand dance of heavenly bodies visible from such a place! And all of these possibilities rest in the glimmer of just one tiny flicker of light on a summer evening.

But there are more than dreams—more than even these possibly true dreams—riding on the wings of Cygnus. There is, beyond the beauty of Deneb and the stars of the pattern, further visual loveliness for the unaided eye. Most prominent is the great Cygnus Star Cloud, one of the brightest parts of the Milky

Fig. 47. *The Milky Way in Cygnus, as seen from Cambridge, Massachusetts, on September 3, 1975. The brightest star is Deneb, and nearby is the North American Nebula. Photographed by Dennis Milon with a five-minute exposure on Fujichrome at ASA 200 and a 45-mm, f/1.8 lens.*

Way, shining like a misty wreath about the extended length of the Swan's neck. Another bright region, just to the east of Deneb, is the North American Nebula. Its precise shape is a surprisingly good replica of the continent, bordered by a Gulf of Mexico and Atlantic Ocean of dark nebulosity, but it is only seen in detail with optical aid and in photographs. The rough form of the nebula, however, is quite prominent to the naked eye in clear, dark skies. The North American Nebula seems to be about the same distance as Deneb and it is possible (though recently disputed) that the star is at least one of those responsible for illuminating the nebula—though the actual space separation between the two is probably at least 70 light-years!

Even more interesting than the North American Nebula is an object that is harder to see. An experienced observer can sometimes spot it with binoculars, but usually a telescope is required. The object (or, really, set of objects) consists of several luminous clouds of gas which are known collectively as

Fig. 48. *Western loop of the Veil Nebula in Cygnus. Photographed with the 100-inch telescope.*

the Bridal Veil or (more often) the Veil Nebula. The clouds do look like diaphanous veils or scarves of luminosity in a telescope. What is most interesting about the Veil is its origin: these glowing clouds are the remains of a supernova, the most violent variety of exploding star. Earlier in our journey we passed the Crab Nebula in Taurus, the expanding remnant of a supernova which occurred less than a thousand years past—but the star which caused the now several-degree-long strands and wisps of the Veil Nebula probably exploded about 30 to 40 thousand years ago. How bright was this star under a Swan wing tip when it shined in the skies of our distant ancestors? In recent years, we have learned a lot more about historical supernovas and some of the staggeringly beautiful (but also, rarely, dangerous) possibilities. Whatever the original Veil supernova may have looked like, the current nebula is elusive and lovely in telescopes. And in photographs, its red, white, and blue silky light-strands—wisps of a continuing explosion *seemingly* frozen for us by vast distance and time—may be the most delicately lovely image given to us by the astronomers.

Other objects of interest in Cygnus include a recent (1975) nova which briefly approached Deneb in brightness; a variable star whose brightness changes by around 9 magnitudes over a period of about 406 days: and the most likely candidate for a black hole. But the final note on Cygnus here concerns its yearly swan song: each year Cygnus approaches the sun and at last sets with it, disappearing from the evening sky. In this fall into the west each year—or over the course of a night—the swan pattern's orientation to the horizon changes so that it seems to swirl about as it descends. The final position in this celestial ballet has the Northern Cross standing perfectly upright on the western horizon! Does this sight which occurs in early evening as autumn wanes therefore stand as memorial to the end of the growing season, or the year?

But we have only just reached autumn in our tour, and the Summer Triangle is visible in the west for early-evening viewers long after summer is gone. Within and near the giant Triangle also lie several small but pretty constellations. Among them is tiny Sagitta the Arrow, flying between the Eagle and the Swan. There is also delicate little Delphinus the Dolphin, shaped like a diamond with a tail and also known, unofficially, as Job's Coffin—an intriguing title, now of unknown origin and explanation. The Dolphin is outside of the Summer Triangle and just east of the Milky Way—it is the westernmost constellation in an area of the sky which is a dramatic contrast to the brilliant Milky Way regions we have just surveyed. As autumn comes and Cygnus twirls into the west, even the Milky Way runs more dimly for a while after it leaves the Swan. But in the northeast there are at least a few bright constellations. In the southern sky, where Delphinus swims, is the largest, dimmest area of the heavens: a mighty sky-spanning realm called the Celestial Sea or, more simply, the Water.

The Water derives its name and defines its boundaries from the fact that all the constellations here represent things aquatic. The most certain members of the Water are Delphinus (Dolphin); Capricornus (Sea-Goat—half goat, half fish); Aquarius (Water-bearer); Piscis Austrinus (Southern Fish); Pisces (Fishes); Cetus (Whale). All of these except Delphinus are huge constellations and to them may perhaps be added, on the borders, Eridanus the River and Grus the Crane (it is also true that two more constellations, Pegasus and Taurus, are often depicted without the back halves of their bodies, which are sometimes said to be submerged in water, out of which they are leaping). What is most interesting is that every one of the main constellations of the Water is faint on the whole, so that this vast area really does bring to mind the great dark expanse of an ocean—indeed, of the vastest ocean of all, space. And darkness, flecked only with a few somber stars, is appropriate to this poignant season of lengthening nights, deepening cold, and departing leaves and birds: the year's demise in the northern world.

There is certainly beauty in the expanding darkness and departure of things (which will, we must remember, be returned and renewed in the growing light of spring). And although the constellations of the Water are dim, they are for the most part interesting. Capricornus the Sea-Goat looks more, as Guy Ottewell has said, like a boat. At its pointed prow (on the west end) is Beta Capricorni and, very near, Alpha Capricorni, which is a double star whose components can be distinguished with the naked eye. Capricornus is one of the most ancient of constellations: the Sea-Goat (front half goat, back half fish) represents the Sumerian god Ea Oannes, who was possibly popular as early as 4000 B.C. and has rich connections with Eden, the Flood, and a maybe full-moon-bright supernova!

Capricornus is found in the zodiac east of Sagittarius, the most southerly point, and is in turn followed by Aquarius the Water-bearer. Streams of faint stars seem to flow out from the center of Aquarius, this center being an asterism called the Urn or Water Jar. These rivulets are supposed to be running down to the mouth of Piscis Austrinus, the Southern Fish. The mouth of the mostly faint fish is marked by what one great amateur astronomer, Leslie Peltier, has called the Autumn Star, to indicate its singular significance. The Autumn Star is Fomalhaut, the only first-magnitude star among the autumn constellations, the only one visible across an entire third of the heavens east to west. Fomalhaut shines at magnitude 1.17 (almost exactly the same as Pollux) but it stands out as a star of greater prominence from the lack of competition. It is like a solitary lighthouse shining near the southern shore of the Celestial Sea. White Fomalhaut lies at a distance of about 23 light-years.

Fomalhaut is the most southerly first-magnitude star visible from 40°N, and so for observers in the United States it is always low, often hidden by trees or buildings. The easiest way to locate it in such a situation, as well as several of the dim constellations here, is by the landmark (skymark?) of autumn, the Great Square of Pegasus. This rough square of second-magnitude stars lies just above the Water, high in the south on October and November evenings. The Great Square is the body of Pegasus, the famous Winged Horse of Greek mythology. The pattern of the entire constellation bears a good resemblance to a horse, except that it is, like Hercules, now upside down in the sky for United States and European viewers. The neck of Pegasus therefore extends to the southwest where his nose is represented by the bright second-magnitude Enif. The front legs of this legendary creature are formed by stars that are mostly dim; the rear legs are bright, but these stars officially belong to another famous constellation, which we will be visiting soon. The Great Square itself is composed, in clockwise

PERSEUS
CASSIOPEIA
CAMELOPARDALIS
AURIGA
ORION
TAURUS
ARIES
TRIANGULUM
ANDROMEDA
PISCES
CETUS
ERIDANUS
FORNAX
SCULPTOR
PHOENIX
CEPHEUS
URSA MINOR
Capella
Menkalinan
Aldebaran
The Hyades
The Pleiades
Mirfak
Algol
Double Cluster
Polaris
North Celestial Pole
Little Dipper
Errai
Alfirk
Rukbah
Tsih
Shedir
Caph
Almak
Great Galaxy M31
Galaxy M33
Mirach
Hamal
Sheratan
Mesarthim
Sirrah or Alpheratz
Scheat
Great Square of Pegasus
Algenib
The Circlet
"First Point of Aries"
Celestial Equator
Ecliptic or Road of the Sun
Menkar
Kaffaljidhm
Risha
Mira
Azha
Diphda
South Galactic Pole
Ankaa
Crab Nebula
Galactic Anticenter
The Kids
Nath
Spiral Galaxy M81
The Pointers
Dubhe
Merak
Phecda
Megrez
Alioth
Mizar
Alcor
Big Dipper
Owl Nebula M97
Pherkad
Kokab
Rana
Zaurac

Fig. 49. *The starry sky for around 10 P.M. (daylight saving time) in October. Derived from a master chart by Guy Ottewell. © July 1975 by Guy Ottewell.*

order from the northwest corner, of Scheat, Markab, Algenib, and Alpheratz. Alpheratz is the brightest of the four but is the one which actually is owned by the neighbor of Pegasus. A line drawn down through Scheat and Markab—the western side of the Square—eventually leads almost directly to Fomalhaut. Directly beneath the middle of the Square are five stars called the Circlet, a small wheel of third- and fourth-magnitude sparklers. The Circlet is the head of one of the two fishes which are Pisces the Fishes, the zodiac constellation following Aquarius. Dim Pisces is pictured as this western fish beneath the Great Square, tied by cords to another fish, which faces north near to the east side of the Great Square. The star Rischa is the "knot" (literal translation of the name) holding together the two cords to which the fishes' tails are tied.

To the southeast of Pisces floats Cetus the Whale. But the pattern seems to be a creature with a long thin neck—more like a Loch Ness monster of sorts. Although "Cetus" means "whale," this pattern has, in fact, usually been identified with a sea monster of unspecified form who plays a role in one of the fullest and most rewarding tales of Greek mythology. Almost all the main characters in that story are represented by constellations in the autumn sky, as we shall see presently. But this strange Cetus is also very interesting astronomically. On either end of the long creature are rather bright stars. The western extremity is Diphda (sometimes called Deneb Kaitos—"the tail of the whale"), a bright second-magnitude star which beams in the southern void across from lonely Fomalhaut. The only other second-magnitude star among the main constellations of the Water is Menkar, in the rather prominent configuration of stars which forms the head of Cetus.

But the most remarkable star in Cetus is one which is not even visible to the naked eye most of the time. It is a variable star whose minimum brightness is usually about ninth-magnitude. But if your naked eye discovers no sign of a bright star here in the long neck of the Whale, look again in a month or two. You will quite possibly find that the position is now occupied by a star of the fourth or third magnitude—or even brighter. This star's name describes it well: Mira—"the Wonderful." It was first recorded in Europe by David Fabricius in

1596 and 1609, though there are Chinese records which seem to refer to it earlier. Like Betelgeuse, Mira is a long-period variable star and red giant, but the variations of Mira's brightness are usually of at least 6 magnitudes in its average period of 331 days. That is an increase of about 250 times, and few stars which are not novas have such a range. Chi Cygni, in the neck of another long-necked creature, the Swan, has an even larger range (8 magnitudes and more!) but only reaches about 5.3 (quite dim) at an average maximum. Mira, however, has frequently reached second magnitude, and on one occasion—in 1779—William Herschel estimated it as almost the equal of Aldebaran! Mira is thrilling to follow to its maximum. I well remember several of the fine maxima of the late 1970s, when Mira rivaled Diphda. Estimate this star's brightness every clear night around maximum, comparing it carefully to nonvariable stars in the area.

The Wonder of the Whale, Mira Ceti, is especially welcome as an occasional bright addition to the dim Water, but providing a bright frame to the celestial dark in the south and the higher Great Square are constellations climbing high in the north and overhead. The chief of these are the human characters in the great myth which includes Cetus and Pegasus. Only the monstrous Cetus is separated from the large contiguous area which contains Cepheus, Cassiopeia, Perseus, Pegasus and—one of the most famous names in astronomy—Andromeda. From Cygnus sinking into the northwest we can follow the dwindling Milky Way through an area of dimmer stars (such as meager little Lacerta the Lizard) in the northern sky, until we come to Cepheus the King. Cepheus, whose shape is like a little building or perhaps a throne, contains an important regular variable, Delta Cephei (this star increases from 4.3 to 3.6 in about 1½ days, with an entire period of 5.37 days). Also here is one of the reddest stars in the sky, Mu Cephei, often called Herschel's Garnet Star (this star, too, is variable but even at its brightest, about 3.7, good binoculars or a small telescope is required to appreciate the very beautiful hue). Beside a King should be his Queen, and so it is in the heavens: next to Cepheus on the Milky Way is Cassiopeia.

According to one of the great Greek myths, Queen Cassiopeia bragged that she was more beautiful than the nymphs of the sea, who were so angered they asked the gods for vengeance. In response, a terrifying sea monster was sent to ravage the coasts of the country until the king and queen should agree to sacrifice their own daughter to it. Cepheus and Cassiopeia were forced by their subjects to comply and so their innocent lovely daughter, Andromeda, was chained to a cliff by the sea—to await the monster. While she waited hopelessly, dramatic events were taking place far out at sea. On an island out there lived the three sisters who had once been beautiful but had boasted so shamelessly that they were turned into the most hideous of beings, the Gorgons. The worst of them was Medusa, who had snakes in place of hair and, like her sisters, was so horrible that anyone who looked at her was literally petrified—turned to stone. Of the three, only Medusa was mortal, and because she and her sisters were such a danger to unknowing travelers, the gods looked for a hero brave enough to try to kill her. At last Athena found the young Perseus, to whom she gave Hermes' winged sandals, a cloak of invisibility, and some other useful gifts. Perseus flew to the island and saw Medusa's distorted (and therefore safe) reflection in his brightly polished shield. Watching only the image, he swooped in, caught her unaware and chopped off her head of hissing locks. From the blood of Medusa, it is said, the beautiful winged horse Pegasus leaped (later to be tamed briefly by another hero, Bellerophon).

But meanwhile, back at the sea cliff, the tide was coming in and the monster had appeared. Just when it seemed that the giant beast would reach helpless Andromeda, down flew Perseus. He drew the monster's attention and got it to chase him until they were far enough from shore—then, he pulled something out of a bag and held it up. It was Medusa's head! The sea monster, jaws agape, instantly hardened into rock (some say coral) and sank senseless to the bottom of the sea, becoming an island. Perseus, of course, then freed Andromeda, won her hand in marriage, and lived happily ever after with her, ruling their own kingdom with wisdom.

But according to star lore, vain Cassiopeia was further punished, for when she was sent to the heavens with all the other participants in this story, she was bound to her throne and made into a north circumpolar constellation (like the two Dippers or Bears and Draco) so that some of the time she hangs upside down most uncomfortably. Of course, inverted posture is the fate of other constellation figures (for various reasons), including Cepheus, who is circumpolar himself.

But the brilliant stars of Cassiopeia do form a very convincing chair or throne, and the constellation is perhaps as beautiful as the Queen imagined herself to be. Cassiopeia is second in prominence only to the Big Dipper among the north circumpolar constellations, those that never set as seen by observers in northern lands. Interestingly, Cassiopeia is almost directly opposite the Big Dipper in the circle they make around the north celestial pole (and both are patterns about equally far from the pole and its North Star). Therefore, in spring, Ursa Major and the Big Dipper ride high in the early evening, but in the autumn it is Cassiopeia that is far above. The constellation's pattern suggests things other than a chair, or course: Cassiopeia is probably Wilwarin, the Butterfly, in the modern mythos of J. R. R. Tolkien. The pattern of Cassiopeia was temporarily altered in 1572, when the Queen was the showcase for a "guest star"—in this case, a supernova brighter than Venus and often named after the famous astronomer who observed it carefully: Tycho's Star. Unlike the supernovas which produced the Crab and Veil Nebula, it has left no visible trace of its explosion that could be found—though a powerful source of radio energy in about the right place may be the remnant.

The Milky Way, having reached its most northerly point in Cassiopeia and Cepheus, turns southeast to prominent Perseus. The brightest star here is second-magnitude Mirfak, about which is gathered a beautiful collection of celestial gems that is superb to the naked eye, and best of all in binoculars or "rich-field telescope" (the group covers too large an area to fit into the field-of-view of an ordinary telescope). This group is probably a true cluster but has been strangely neglected by amateur astronomers for a long time. Mirfak and this cluster lie at the center of the roughly K-shaped figure of Perseus, which is a good approximation of a man holding something out at arm's length.

What Perseus is holding out is Medusa's head, and the star which represents it could not be more appropriate. It is Algol—"the ghoul"—a variable star whose striking behavior may have been noticed by and disquieting to the ancient Greek and medieval Arabic observers. While Mira is the most remarkable example of a long-period variable, Algol is the brightest of another class of variable stars, called *eclipsing binaries*. In an eclipsing binary system two stars revolve around each other (actually around a barycenter) and we see them at an angle such that at least one of the stars periodically hides (or partly hides) the other. In the case of Algol (Beta Persei) the brighter star gets mostly eclipsed by the fainter so that the magnitude, usually 2.3, drops swiftly down to 3.4 once every 2.86739 days. In between every two of these deep dimmings, there is a very slight one which occurs when some of the fainter star is hidden behind the brighter. Observe Algol frequently and try to catch it on a night when it is dim (or use the monthly predictions of minima in *Sky and Telescope* magazine; see the Annotated Bibliography). Do not use the moderately bright star beside Algol for comparison: the star is Rho Persei, itself a semiregular variable ranging from about 3.3 to 4.0.

In addition to the Mirfak group, there are several other clusters in Perseus which can be seen with the naked eye. M34 is not very prominent to unaided vision but on good nights can be detected not far above Algol as the constellation rises in the northeast. Far more prominent are the twin objects which lie about halfway between the main pattern of Cassiopeia and the northern end (the helmet) of the Perseus pattern: there is not just one but a pair of magnificent clusters here, the Double Cluster of Perseus. Each cluster by itself would rival any of the other open clusters (even M11 in Scutum) for richness and brightness, but together they make one of the true showpieces of the heavens for telescopic viewers. The cluster Chi is rated as of magnitude 4.7, with about 300 stars (in the telescope), while h Persei is 4.4 with 350 stars. The two have their centers only $\frac{1}{2}°$ apart, so that they can fit together even in the narrow field of a telescope (at rather low magnification). To the unaided eye they appear prominent, and just as they did to the ancients, who knew them as a curious elongated patch in the midst of the richest part of the Milky Way that is north of Cygnus. But on very clear nights (and with knowledge of what this glowing spot really is) the naked eye distinguishes the roughly spherical forms of the individual clusters.

The whole area around the Double Cluster is so rich that on superb nights it appears to abound with large bunchings of stars, whether you look with eye, binoculars, or a rich-field telescope. In a telescope, of course, the Double Cluster is awesome: twin concentrated splashes of bright telescopic stars with some beautiful contrasting tints. These mighty clusters are in reality more magnificent than they seem from Earth: they are probably about 7,500 light-years distant, and therefore all their brightest stars must be supergiants—both blue and red ones—with absolute magnitudes comparable to those of Rigel and Betelgeuse. If the Double Cluster were even as close as distant Rigel (900 light-years away) they would be a sight of unbelievable splendor: to the naked eye, each cluster would appear several degrees wide and be composed of many dozens of stars visible to the unaided vision, including a few dozen of zero to first and second magnitude!

Not far from Perseus and Cassiopeia is Andromeda the Chained Maiden. She is formed by two lines of stars running from (and officially including) Alpheratz of the Great Square of Pegasus. The southern line is by far the brighter, consisting of Alpheratz and two more stars which are almost as bright. These three bright stars are almost equally spaced along the line, and Mirach and Almach (working out from the Square) are respectively reddish and orange. Almach is actually a fine double star in telescopes, often compared with Albireo in Cygnus. Despite all these facts, the dim northern line of Andromeda is the more interesting by far, simply because there lies near it an object about as prominent to the naked eye as the Double Cluster—and yet unimaginably more splendid than even those breathtaking star-cities. It is the Great Andromeda Galaxy, which we will encounter in the next chapter along with our own galaxy and the M33 galaxy in the constellation of Triangulum.

Triangulum the Triangle and Aries the Ram are two small but rather conspicuous constellations below Andromeda but still high in the southern sky. Aries contains the 2.0-magnitude orange Hamal, but the constellation's fame is as a member of the zodiac. The Ram is traditionally the first constellation of the zodiac because it was once the place where the ecliptic crossed the celestial equator—the location of the sun at the vernal equinox, the first day of spring. And the first day of spring was the beginning of the year in some lands, including

Fig. 50. *The Pleiades photographed with a five-inch Cogshall camera at Lowell Observatory. Note the prominence of the Merope Nebula (around and below the bottom star of the major ones in this view). Reprinted from* Burnham's Celestial Handbook *by courtesy of Robert Burnham, Jr.; Lowell Observatory photograph (by permission).*

England until as late as 1752. Now precession has shifted the Earth's axis so that the place where ecliptic crosses equator, and spring starts, is farther west, in Pisces. But tradition is often strong, and this spot is still called the First Point of Aries. The vernal equinox will move into Aquarius in A.D. 2597, initiating the so-called "Age of Aquarius."

The vernal-equinox position in the sky forms part of the basis for astronomers' system of celestial coordinates. We have already dealt with the north and south celestial poles and celestial equator. These are at +90° (90° north), −90° (90° south), and 0° *declination*—the measurement which is on the celestial sphere of the sky equivalent to latitude on the sphere of Earth. But just as on Earth we need a prime meridian for longitude, so too there must be a starting point for measuring west to east on the celestial sphere. That measurement is called *right ascension* (*R.A.*) and is usually figured not in 360° (as longitude on Earth is) but rather, for greater usefulness, in hours, minutes, and seconds—a measurement of distance around the heavens. The prime meridian of the sky, the 0-hour (0h) line of R.A., passes almost down the east side of the Great Square of Pegasus and through the vernal-equinox point in Pisces. Betelgeuse is at nearly 6h, Leo's tail near 12h, the Teapot's spout (and center of our galaxy) near 18h, and when we pass 23h and reach the next hour of right ascension we have completed 24 hours and have arrived back at 0h, the prime meridian. The R.A. and declination of other stars and constellations we have visited in our tour can be measured on the star charts in these past two chapters.

On the other hand, thousands of years ago the vernal equinox had not yet come as far west as Aries, and resided in Taurus the Bull. According to Guy Ottewell, it may not be a coincidence that our letter A resembles the pattern of Taurus upside-down: the first letter of the alphabet may have represented the ancient first constellation of the zodiac. Whatever the truth may be, our tour of the heavens has now carried us from first to last through the year: we have returned to Taurus the Bull and have just one more celestial object to look at. For here, in the autumnmost part of the winter Bull, shine the Pleiades.

The Pleiades or Seven Sisters make up the most beautiful

star cluster in the heavens. Only a few other clusters contain individual stars visible to the naked eye, but these are generally fainter or much more loosely scattered. Actually not seven but six of the Pleiades stars are easy to see with the naked eye; at least five or six more are bright enough but require good eyes or good skies to see individually because they are very near to the main members of the cluster. These main stars form a rough and very tiny dipper, so that some beginning stargazers have thought the Pleiades were the Little Dipper (which is actually much larger and much less conspicuous). It was the ancient Greeks who called the Pleiades the Seven Sisters, daughters of Pleione and the giant Atlas who supported the sky on his shoulders. Many scholars have wondered if the Greeks could really see seven stars here in a casual glance (their night skies were certainly cleaner and darker than ours) or whether they just added the seventh because seven was a mystical number. Perhaps one of the Pleiades has become fainter since ancient times? We do not know (the star Pleione is slightly variable in modern times) but there are many legends about the Lost Pleiad in cultures as widely scattered as those of Japan, the Gold Coast of Africa, Borneo, and Australia. The story of a lost Pleiad also appears in Mongolia, where it is said that seven robbers (the seven bright stars of the Big Dipper) kidnapped one of the seven maidens of this cluster and carried her far off to the north where she now shines as Huitung Ot, "Cold Star," or Alcor. Alfred Austin has written: "The Sister Stars that once were seven / Mourn for their missing mate in Heaven."

The "handle" of the Pleiades dipper is formed by two close-together stars which have been named for the Sisters' parents. The stars in and around the "bowl" have received names of the daughters, the brightest of these (and brightest of the cluster) being the 2.9-magnitude Alcyone. Though not so rich as such open clusters as the Double Cluster in Perseus, the Pleiades are much brighter (in apparent magnitude). The combined magnitude of the main stars and the many which are only visible in the glorious view through binoculars amounts to a brightness approximately equaling that of Regulus. A touch of nebulosity is visible in telescopes in dark skies (and, theoretically, just barely to the naked eye) but long-exposure photographs show that the Pleiad stars are splendidly shrouded in thick, glowing nebulosity, and we know that the Pleiades are young hot blue stars, at a somewhat later stage of development than the awesome Orion Nebula.

The Pleiades are possibly more myth-honored than any other star group or constellation. Almost every culture has told stories about them, and in many cases they have been pictured in gentle terms and figures, often feminine. One tribe in Brazil has based its entire calendar on the rising and setting times of the cluster. The Druids of ancient Britain held their rites from which we eventually got Halloween when the Pleiades started rising in the early evening at the start of November. One American Indian legend says that seven maidens were chased up a mountain by two enormous bears whose claws on the mountain's sides turned it into the peculiar formation we now call Devils Tower (Wyoming). But the maidens were saved by being lifted up to heaven to become the Pleiades. Did Steven Spielberg know that the people in his movie *Close Encounters of the Third Kind* were not the first to use this peak as a launch point for a voyage to the stars?

The first recorded observation of the Pleiades is a Chinese one dated 2357 B.C.! Since then, not only legends but great works of literature have praised the beauty of these stars. The experience of seeing the lovely sisters twinkling together in an intermingling of star beams was well expressed by the poet Tennyson in "Locksley Hall": "Many a night I saw the Pleiads, rising thro' the mellow shade, / Glitter like a swarm of fireflies tangled in a silver braid." But perhaps the most chillingly beautiful reference to the Pleiades is in the Bible itself. It is in the Book of Job that God makes sharp-as-lightning clear the difference between God and man in an awesome roll of questions to Job, one of which is: "Canst thou bind the sweet influences of Pleiades, or loose the bands of Orion?"

A hauntingly lovely legend about the Pleiades is told by the Hervey Islanders of Polynesia, who say that the cluster was once a single star, brightest in the heavens, as radiant as a half-moon. The god Tane overheard the great Pleiad-star boasting shamelessly of its beauty and became so angry that he enlisted

Fig. 51. *The Pleiades photographed with the famous 13-inch astrographic telescope at Lowell Observatory. Reprinted from* Burnham's Celestial Handbook *by courtesy of Robert Burnham, Jr.; Lowell Observatory photograph (by permission).*

Fig. 52. *The Pleiades. Photograph by Robert Burnham, Jr. Reprinted from* Burnham's Celestial Handbook *by courtesy of the author.*

the aid of Mere (Sirius) and Aumea (Aldebaran) to punish the star. One night, when the heavens were thickly clouded over, the confederates crept up, but the great star heard them coming and fled. First he hid under the stream of the Milky Way—but

Mere diverted the flow, revealing him. Then the chase was on again but the Pleiad-star sped away so swiftly that it was obvious he would soon lose his pursuers. But at that moment Tane suddenly seized Aumea and hurled him with such force that Aumea struck the Pleiad-star, shattering the great light into six pieces. Tane was satisfied, Mere was pleased indeed to be the new brightest of all stars, Aumea was happy to have his own light no longer overwhelmed by the nearby beauty. The six little stars, thereafter called Tauono, went forlornly back to their original position. But it is said that the little pieces still look longingly at their reflection when they pass over quiet waters, and still find themselves beautiful—as we do.

As the autumn night grows late, the Pleiades are rising higher and below them the Hyades, Aldebaran, and the rest of Taurus. And now at last the grandest sight among all the constellations: the enormous figure of Orion, mighty Hunter, tilted, looms on the eastern horizon. Winter is coming and with it again the year's most splendid congregation of bright stars.

Our year and our journey are done—and yet the journey never really ends. Each year of our lives we can return and find among the stars new sights, new tales, new facts—and everlasting beauty to inspire and uphold us. And each year, the old stars and constellations we already know will become for us still more familiar and loved old friends.

CHAPTER TEN

The Milky Way and Beyond

Our last setting in this book is an evening in early summer. One of the longest days of the year has ended, and long twilight is finally giving way to night. Yesterday's heat and unstable atmosphere engendered lines of violent thunderstorms whose vanguard was a tumbling black roll cloud in the late afternoon. All night long heaven and earth were battered and strummed, seared and snapped, made the resounding mouth and throat of hoarse and melodious bellows, alive, laced, lanced, and leaped over with instant-flowing and many-branched lightning, drenched with a dancing spatter and tumult of rains. Today all was cool and refreshed, and now, as darkness falls at last, all is calm. Fireflies here and there express a great peace, yet somehow also beckon. And unimaginably far above them—or maybe just a dream's length away—the overspreading night sky itself is becoming ever more thickly blanketed with the fireflies' distant relatives, the stars. The stars also flicker, but never quite go out, and amongst them, in the east, curves the course of a river of soft radiance, the Milky Way, the river of stars.

The Milky Way is the galaxy we live in, and the photographs of astronomers show us that from the outside other galaxies seem less like rivers than like clouds or ponds, or, most beautiful of all, vast whirlpools or hurricanes of stars. These are, respectively, the irregular, the elliptical, and the spiral galaxies. The true forms, motions, and compositions of these things are in some respects astonishingly unique, but the analogies hold some considerable truth because each of these mighty systems contains stars as numerous as droplets in a cloud, and the motion in them is most like that in a fluid or vapor, though the wind that blows (or current that flows) through them, giving them structure, is the gravity that arises from the very existence of their mass. Those lovely spiral galaxies spin round like hurricanes with arms of gas, dust, and stars, curving exquisitely out from a thicker hub of most intense starlight, the often multi-billion-sun central hub of the galaxy. Spiral galaxies are strange storms of stars, of light, of splendor, but in looking at a gallery of their photographs we are struck by the varying degrees to which their starry arms (what would be lines of thunderstorms in a hurricane) are tightly or loosely coiled. Could these views be snapshots of different stages in the evolution of a spiral galaxy? There seem to be transitional forms between the spiral, elliptical, and irregular galaxies. Could they all be stages in a galactic life and growth? One is reminded of another, and far more humble, Earthform, the flower, which many galaxies somewhat resemble. Perhaps galaxies too go from bud to blossom and maybe even to seed?

As we think of these fancies, we can smell blooming flowers around us in the still air of this early summer evening. But for the time being, we place in the back of our mind this persistent, evocative scent along with the vision of galaxy flowers strewn across the fields of space. We gaze high in the east at that band of luminosity and wonder how this sight we call the Milky Way would fit into a photograph of our Milky Way galaxy taken from afar.

The Milky Way galaxy is a spiral. Like most whirlpools—such as that whirlpool of the air, the hurricane—a spiral galaxy is far wider than it is deep. Far more so than a typical child's pinwheel, these galaxies are flattened, with most of the material outside of the round central hub confined to an equatorial plane, so that from a side view the galaxy looks roughly lens shaped. An interesting departure from the lens structure is the globular star clusters, those balls of hundreds of thousands (or more) stars which we encountered in the last chapter and which are each somewhat like miniature central hubs. The globular clusters form a spherical halo surrounding the hub itself, in some cases at great distances. The position of our own solar system in all this majesty is near to the equatorial plane, in the

Fig. 53. *Two globular star clusters (NGC 6522 and NGC 6528) near the center of our galaxy in Sagittarius. KPNO four-meter photograph.*

equatorial disk, but well over halfway out from the center. Of course there is no abrupt edge to our galaxy, but according to estimates based on one possible set of criteria, the Milky Way galaxy is about 100,000 light-years across, with our position about 30,000 light-years from the center. Thus we live in a rather rural area, if not quite the galactic outback! The reason why the Milky Way looks like an arching band in summer and then again, but more dimly, in winter is that we are, in those sections of the sky, looking through the densely populated equatorial disk in which we ourselves are located. The band is dimmer in winter because we are then looking outward to the edge of our star-storm of a galaxy, through part of our own spiral arm and at an outer arm. But on a summer night, like the one on which we now stand outside, we look inward along our own arm and through inner ones toward the center of the galaxy in Sagittarius. We cannot see the densely star-populated center itself, for it is hidden behind clouds of dust and gas (both bright and dark), but those clouds and the star clouds which are just outliers of the central magnificence are still the most glorious section of our heavens from Earth!

In winter the Milky Way band trickles through its narrowest and faintest section in Auriga after having reached its most northerly point in Cassiopeia and Cepheus and intensifying in spots like the area of the Double Cluster in Perseus (we are looking at the Perseus arm of the galaxy, outward from our own, when we gaze here). After Auriga, the Milky Way crosses the ecliptic at a point which, by an interesting coincidence, happens to be where the sun is at summer solstice—the meeting point of Auriga, Gemini, Orion, and Taurus. The Milky Way is rich in open clusters, a few visible to the naked eye, as it flows on down behind Orion and between the Dogs, and is soon after lost to the horizon for viewers as far north as the United States. But tonight it is summer and Cassiopeia and Cepheus are still low in the northeast from where we can follow the band southward down the summer side of its circle, the more glorious one, the one which will bring us at last to the region in whose direction lies the mighty center of our galaxy.

High in the northeast above Cepheus now, and destined to be overhead at a later hour or a later month, we can see the Cygnus Star Cloud as a bright star-mist located in the midst of the most beautiful of constellation birds. But the flight here is not just the mythical one of an imagined swan. The stars of our galaxy are in flight around the center in orbits which must change somewhat in size and shape eventually because of the fluid-like redistribution of matter in the galaxy—and for stars as far out as the sun, the orbits may be never twice the same because each circuit takes so long to complete. When we gaze toward bright Deneb, the tail of Cygnus, we are looking in the general direction of our future for it is that way that the stars in our neighborhood are collectively headed. The sun's individual course varies somewhat from the mean of these stars and therefore *seems* when measured against them to be headed for a point in Hercules not very far from the summer sapphire Vega. The point is called the Apex of the Sun's Way (the Antapex or Sun's Quit is located well to the south of Orion in the constellation Columba the Dove). If we compensate for the various effects of perspective we find that the sun's true destination is only slightly different from the mean destination for local stars: it is believed that we are heading for a point not far from Deneb and that fascinating 61 Cygni star system. The stars which lie in this direction will, of course, have themselves traveled on by the time we have reached that point—we use them only as a reference to orient ourselves on the galactic level in the ever-shifting but regular system of things. But it is very encouraging to see our pathway into the future lit before us with such beautiful stars; and it is both enchanting and awesome to sit out on a summer evening and behold where all of us—not just humankind but every plant, animal, stream, and stone, even the moon and sun and many of our naked-eye stars—where all of us are going to. Everyone and everything in the universe, it seems, is a voyager.

One of the very few traveler stars visible to the naked eye which are not going our way is now shining high in the west (facing east-crowning Vega) at this hour. It is our friend of a few chapters back, orange Arcturus, and in the midst of our current thoughts we can appreciate far better the distinction of this gypsy star, wandering on its lonely way, swooping down through the galaxy's equatorial disk in no more time than it takes for one species to come to fruition—or at least the first fruits of that species' great labors, if our civilization can endure a little longer to find its wisdom. But Arcturus, like the sun, is yet an orbiter of the galaxy, and all lengths of time we have so far talked about in this book are very brief compared to the orbital period of a star at the sun's distance from the galactic center. The period of the "galactic year" (as we might call it) for our solar system is believed to be about 200 to 250 million years (revolutions of Earth around the sun). So the full reign of the dinosaurs probably lasted much less than one galactic circuit. And even the sun and Earth have probably only existed 20 to 30 galactic years—an interesting coincidence considering that our many-ages-old world and sun are indeed likely to be young adults in the society of the Milky Way galaxy.

Between the two birds Cygnus and Aquila, and their bright stars Deneb and Altair, a vast tongue of darkness splits the stream of light. It is the Rift—not an absence of stars, but thick clouds of obscuring dust which are situated along the equatorial plane. The lesser channel of the Milky Way band flows on for some distance before apparently trickling into nothingness. On very clear nights try to determine how far you can detect this huge "tributary," as well as other boundaries of the Milky Way band. How close to the tiny main pattern of Lyra does Milky Way luminosity extend? Notice also on some nights how sharply defined areas of different intensity appear, and the prominence of dark areas like the Rift (notice the dark promontory which cuts far into the Milky Way a little way northward from Deneb). On extremely clear country nights an astonishing amount and detailed wealth of Milky Way can be seen. The veteran and gifted observer Walter Scott Houston writes of some nights on which he has seen "wisps and patches" of Milky Way extending far into the constellation Andromeda, even as far as a famous object in that constellation which we will encounter in a moment. It also seems to be true that on hazy nights the Milky Way is more readily visible than individual faint stars. Test this claim for yourself and see if you can figure out why it might be so.

Bright beside the widest part of the dark Rift is the Scutum Star Cloud, a wondrous patch midway down the southern sky for United States viewers. That radiant cloud almost overwhelms the naked-eye view of the great Scutum star cluster, M11, but there is yet greater glory waiting farther south. For it is now that the Milky Way band swells to its greatest width and prominence of all. When we get a clear view of the lower part of our southern sky it is sometimes at first difficult to believe that we are not looking at giant clouds in Earth's atmosphere somehow illuminated. But the clouds here are not composed of water droplets, they are composed of star after star, thick stellar concentrations which are part of the wall between us and the center of the galaxy. If the Milky Way band is imagined to be a river then this Sagittarius region is a mighty delta, bright with broad floodplains of soft radiance. The river passes down to the southern horizon, feeding some unimaginable ocean of time and space—or timelessness and infinity—beyond our view.

Fig. 54. *The figure of a young person is seen silhouetted against, and reaching for, the star clouds of Sagittarius, which guard the heart of our galaxy. Photograph by Robert Burnham, Jr. Reprinted from* Burnham's Celestial Handbook *by courtesy of the author.*

In the last chapter, we toured the smaller patches of brightness in this ravishingly rich area—among them M24 and the line of nebulas which includes M17 (the Omega Nebula) and M8 (the Lagoon Nebula). But we did not locate amongst all these attractions the position about which wheels virtually all of the congregated might of worlds, stars, gas, and dust which our naked eyes can see. On summer nights like this one, I can imagine an interstellar tour ship soaring in from the dim rim of our galaxy on a wave of flight like the edge of a ripple-ring moving across light as if light itself were a still, clear pond. The ship—which might also be Carl Sagan's "night freight to the stars"—hurtles effortlessly through the resistlessness of space, passing our beloved backwater solar system and going down the mainline of the galactic plane past the glows of the Sagittarius nebulas there and onward—all aboard for Milky Way Central! That is the name we could give to the great terminal in the middle of our galaxy. If we went to live on a planet in the very nucleus of the galaxy we would dwell under night skies as filled with radiance as ours seem to be with dark. Maybe hundreds of stars would appear brighter than Sirius does in our sky, and thousands of times more stars be visible to the naked eye—but the background against which they shined would be so brightly filled with countless telescopic stars that the sky would seem clotted with light. It would be a splendid place to visit, but we would miss our dark; perhaps it is better—at least for creatures born on such a world as ours—to live where we can have sight of both light and dark, and know not just the power of one, but the potency and preciousness of both.

There are some astronomers, however, who believe that the very heart of the galaxy, inside of the great ball of close-together stars, is a giant black hole. That would fit well (in a few ways) the analogy of a spiral galaxy to a hurricane, with the black hole as the star-storm's eye. But there is much controversy surrounding this theory. Perhaps our best guess at present is that the Milky Way's heart is bright. We cannot see it directly, our view is obstructed, but when we look just beyond the spout of the Teapot of Sagittarius, where the Archer's drawn arrow points, or at the end of the southbound run of nebulas, we gaze upon the Large Sagittarius Star Cloud (M24 is sometimes called the Small). If our eyes could only pierce through all the clouds, bright and dark, we would at last glimpse it. Our eyes cannot, but at least on a summer evening like this one we can see all the closer wonders of the clusters and clouds of Sagittarius and Scorpius—as we gaze inquisitively, longingly toward the hidden presence of Milky Way Central.

The night is growing very late now. The center of the Milky Way galaxy is setting in the southwest, with the bright summer band trailing through the zenith and more faintly down into the northeast. The darkness of the Celestial Sea floods the entire southern and southeastern heavens. But high in the east Pegasus is galloping effortlessly up the vault of night and behind him is heroine Andromeda, chained lady of myth and star-maiden of the constellations. If we were now in some far southern land we could see two patches of light floating like detached pieces of Milky Way in the sky not very far from the unstarred south celestial pole: the Magellanic Clouds, named for the world's first circumnavigator, Ferdinand Magellan, who was one of the earlier Europeans to sight them. Both of these Clouds are between 150,000 and 200,000 light-years away: the Large Magellanic Cloud is over 20,000 light-years across, the Small Magellanic Cloud about 10,000. Both are usually considered irregular galaxies, but the Large Cloud seems to show traces of incipient spiral structure. The Magellanic Clouds are among the satellite galaxies of the Milky Way; some of the others may be hidden on the opposite side from us, and it is even possible that astronomers have recently detected in the Milky Way the remnants of a small galaxy which collided with us long ago. Such collisions are often like meetings of storms, the larger enmeshing the smaller and blending with it, but it is probably most likely for two colliding galaxies to simply slide through each other, and usually with interesting and profound but not catastrophic alterations. But all of this speculation also slides on through our mind at this time because we are standing on the northern slopes of the Earth and before our eyes a mysterious stain of light shines peacefully to one side of Andromeda's chained right arm. That curious glimmer is no mere fancy, it does not go away: it is the soft flame of about 300 billion stars that are 2.2 million light-years away—the Great Galaxy in Andromeda.

The Andromeda Galaxy, M31, turns out to be a beautiful spiral, probably almost twice as wide as the one we live in. Even at this star-wheel's inconceivably great distance it appears large enough for us to trace its length for several full degrees with the naked eye on excellent nights in very dark skies. The very best visual observations and measurements from photographs suggest that the full length is as much as about 5°. The total apparent brightness of M31 is about fourth-

magnitude, so it is not surprising that this object was noticed long before the invention of the telescope. This galaxy was mentioned by the Persian astronomer Al Sufi before A.D. 986 and by others as far back as A.D. 905. The first known telescopic observation was by Galileo's rival, Simon Marius, who compared the sight of M31 to "the light of a candle shining through horn." The Great Galaxy can be found beside the second star to the northwest of Mirach (Beta Andromedae), which means that observers at about 40°N latitude can see it pass directly overhead. Even if M31 were not lovely to behold—like some cosmic firefly—it would still be very much worth a look because of its nature. The light we see arriving from the Andromeda Galaxy tonight left there over 2 million years ago, when the very species of man was just developing. For many people, this is the farthest thing they will ever get to see with their unaided eyes.

There is, however, for anyone who gets to observe on a clear night far from big cities, one final, probably farther object which is definitely visible to the naked eye. It can be found about as far to the southeast of Mirach as M31 is to that star's northwest, and it is the spiral galaxy M33. Another way to find its faint glow is to look not far to the northwest from the apex of Triangulum the Triangle (the constellation within whose boundaries this galaxy lies). M33 is thought to be 2.4 million light-years away and to constitute, alone with M31, our own Milky Way, and at least 25 other (mostly much smaller) galaxies, the so-called Local Group. The Local Group is an example of a still larger level of organization in the universe, and there may be ever larger levels of order out to as far as we can ascertain. The Local Group itself seems to be a bunching on the edge of a "supergalaxy" whose center is the Virgo Cluster of at least 2,500 galaxies!

We encountered the rumor of these clusters of galaxies a few chapters back when we gazed upon the pretty star cluster of Coma Berenices. Back then we considered the fact that this area is our northern window, when we look up out of the equatorial plane of our galaxy, at a right angle to the Milky Way

Fig. 55. *This excellent photograph of the Great Galaxy in Andromeda, M31, was taken by George East on September 23, 1978, at Mashpee, Massachusetts, with a 15-minute exposure on Fujichrome 100 and a 300-mm lens at f/1.5.*

Fig. 56. *The Whirlpool Galaxy, M51, in Canes Venatici. The satellite galaxy at the end of the spiral-arm extension in this rather strange system is NGC 5195. Photographed with the 200-inch Hale telescope.*

band. If this is our spring "skylight" to look out of, then our house must be the entire Milky Way galaxy, small against the sky of the greater universe. In autumn there is a window in the southern sky, looking south out of the galaxy, with M31 and M33 near the window's edge. And these are close neighbors of ours. The Coma cluster of galaxies is centered at least 100 times as far away as M31, and we see the farthest galaxies and (as yet unexplained) quasars as they were at a far earlier stage in the entire life of the universe—for some of these have been detected over 10 billion light-years distant. The photographs astronomers take of these objects show us the way these things looked before there was even a sun or Earth in existence.

This book is not the place for discussion of the many speculative ideas about the structure and development of the physical universe, the proper concern of cosmology. But it is well known even to laymen that the leading scientific theory about the origin of the universe is the Big Bang—evidence for which is the fact that galaxies show shifts in the lines of their chemical spectra which may mean that these objects are all receding from us and receding the faster the farther away they are. If that is true, it could mean that all galaxies or groups of galaxies are hurtling away from each other in an expanding universe which many billions of years ago was concentrated in one stupendously dense mass that exploded with the great birth pang of all. But is this all there is to the story of our universe? There seem to be some people who think so: who believe that we have only a few loose ends to clear up, a few points to elaborate on. Are the astronomers' cosmological theories hardening into deathly realities, do we have little left to learn? Have we almost reached, as some scientists have written, the solution to the universe's deepest mysteries? Will the investigations of subatomic physics and astrophysics soon give us "the key to the universe"?

It seems to me very clear—and also very cloudy and tangled, both dark and light—that this is not so. All clouds and suns, darknesses and lights, all the forever astonishing richness of color, form, and motion in our universe testify to us that the universe we live in retains a variety and a potency of being which exceed our finite attempts to analyze, record, and interrelate its parts. Through every impenetrably solid and irreducibly fundamental formula bursts the secret splendor of things; beyond the grasp of the broadest and most generous philosophy or art flies the further wonder which carries the universe on its way. All generations of thinkers have included those who had the vanity and folly to believe that our latest ideas were bonds constraining existence for our consideration—or for the most crass, materialistic use. Our ideas are sketches. I do not say that these sketches have been hopelessly inadequate depictions of the magnificence of the universe (both physical and spiritual—the latter lying outside the proper province of science itself); I say quite the opposite: these pictures have been a joyous progress of various and ever richer glimpses of the truth. Let us admire them and enjoy them as long as we remember that they are just a lovely glimmer in reflection of the brilliance of full beauty, and a means toward that wonder which is the rightful and boundless end for our deepest appreciation.

If you really do stand out late on a summer night, gaze upon the Andromeda Galaxy while catching the scent of flowers, and then sift through a treasury of facts while perhaps imagining flowering galaxies shot out in a cosmic going-to-seed, never forgetting to keep enough freedom of mind to let plain sight and sense of these things freshly take you, you are then doing what this book has attempted to do. The final—and yet ever ongoing—result of that process of looking and learning is deep enjoyment. The ultimate purpose of observing the phenomena of light (and dark) in the heavens is no mere compiling of data. Human beings are meant for an entire heaven of stirring height and involving depth beyond the recording of so-called fact—a heaven of new perspectives and all the feelings that go with these discoveries of one's intimate participation in the wonder of the universe. By learning about the sky and how to enjoy it, we learn more about ourselves and our role in earth and sky.

In this book we have journeyed from the rainbow in a shower so close that its drops may still be spattering and musically speaking all about us, all the way to that galactic glimmer so distant it began before man and has come to us to meld maiden, constellation, myth, and vast star-system all into a name: Andromeda. We are ultimately and most mysteriously involved with all the wondrous sights in the sky. It is fitting that we may stand right within a rainbow that another observer can see and be part of the fleck of milky light which eyes in Andromeda's galaxy will someday observe. It is our joy to be a part of this wondrous mystery which can carry our eyes and minds freely from the arc of color of the nearest rainbow out maybe even to the shining of the farthest star.

APPENDIXES

APPENDIX 1
The Constellations

M = invented in modern times * = not properly visible from 40°N

Andromeda (the Chained Maiden)
Antlia the (Air) Pump M
Apus (the Bird of Paradise) M *
Aquarius the Water-carrier
Aquila the Eagle
Ara the Altar *
Aries the Ram
Auriga the Charioteer
Boötes the Herdsman
Caelum the Chisel M *
Camelopardalis the Giraffe M
Cancer the Crab
Canes Venatici the Hunting Dogs M
Canis Major the Greater Dog
Canis Minor the Lesser Dog
Capricornus the Sea-Goat (or Goat-fish)
Carina the Keel (of the ship Argo) M *
Cassiopeia (the Queen)
Centaurus the Centaur *
Cepheus (the King)
Cetus the Whale
Chamaeleon the Chameleon M *
Circinus the Pair of Compasses M *
Columba the Dove
Coma Berenices Berenice's Hair
Corona Australis the Southern Crown
Corona Borealis the Northern Crown
Corvus the Crow
Crater the Cup
Crux the (Southern) Cross M *
Cygnus the Swan
Delphinus the Dolphin
Dorado the Goldfish (not Swordfish) M *
Draco the Dragon
Equuleus the Little Horse
Eridanus (the River)
Fornax the Furnace M
Gemini the Twins
Grus the Crane M *
Hercules (the Strongman)
Horologium the Clock M *
Hydra the Water Serpent (female)
Hydrus the Water Serpent (male) M *
Indus the Indian M *
Lacerta the Lizard M
Leo the Lion
Leo Minor the Lesser Lion M
Lepus the Hare
Libra the Scales
Lupus the Wolf *
Lynx the Lynx M
Lyra the Lyre
Mensa the Table (Mountain) M *
Microscopium the Microscope M
Monoceros the Unicorn
Musca the Fly M *
Norma (the [carpenter's] Square) M *
Octans the Octant M *
Ophiuchus the Serpent-bearer
Orion the Hunter
Pavo the Peacock M *
Pegasus (the Winged Horse)
Perseus (the Hero)
Phoenix the Phoenix M *
Pictor (the Easel of) the Painter
Pisces the Fishes
Piscis Austrinus the Southern Fish
Puppis the Poop (of the ship Argo) M
Pyxis the Compass M
Reticulum the Net (observer's reticle) M *
Sagitta the Arrow
Sagittarius the Archer

Scorpius the Scorpion
Sculptor the Sculptor (originally the Sculptor's Studio) M
Scutum the Shield (of John Sobieski, a King of Poland) M
Serpens the Serpent
Sextans the Sextant M
Taurus the Bull
Telescopium the Telescope M *
Triangulum the Triangle
Triangulum Australe the Southern Triangle M *
Tucana the Toucan M *
Ursa Major the Greater Bear
Ursa Minor the Lesser Bear
Vela the Sails (of the ship Argo) M *
Virgo the Virgin
Volans the Flying Fish M *
Vulpecula the Little Fox M

NOTES: The constellation Serpens is composed of two parts, Serpens Caput (the Serpent's Head) and Serpens Cauda (the Serpent's Tail), which are separated from each other by Ophiuchus. The old constellation Argo has, in modern times, been divided up into the now official constellations Carina, Puppis, and Vela (one sequence of Greek-letter designations runs through the three, a holdover from the days when the great ship Argo was still undivided in the sky).

Appendix 2

The Brightest Stars

m = apparent visual magnitude   L = luminosity (sun = 1)
D = distance (light-years)   v = considerable variability
* = not properly visible from 40°N

	m		L	D
Sirius (α Canis Majoris)	−1.42 winter	blue-white	23	8.7
Canopus (α Carinae)	−0.72 winter*	white	1,400	110
Alpha Centauri	−0.27 spring*	yellow	1.5 (A), 0.4 (B)	4.3
Arcturus (α Boötis)	−0.06 spring	orange-yellow	115	37
Vega (α Lyrae)	0.04 summer	blue-white	58	27
Capella (α Aurigae)	0.06 winter	yellow	90 (A), 70 (B)	45
Rigel (β Orionis)	0.14 winter	blue-white	57,000	900
Procyon (α Can. Minoris)	0.35 winter	yellow-white	6	11.3
Achernar (α Eridani)	0.53 autumn*	white	650	120
Beta Centauri	0.66 spring*	blue-white	10,000	490
Betelgeuse (α Orionis)	0.70_v winter	deep orange	7,600–14,000	520
Altair (α Aquilae)	0.77 summer	pale yellow?	9	16
Aldebaran (α Tauri)	0.86 winter	orange	125	68
Acrux (α Crucis)	0.87 spring*	white	3,000 (A), 1,900 (B)	370
Antares (α Scorpii)	0.92_v summer	deep orange	9,000	420
Spica (α Virginis)	1.00 spring	white	2,300	275
Pollux (β Geminorum)	1.16 winter	orange	35	35
Fomalhaut (α Pisc. Aust.)	1.17 autumn	white	14	23
Deneb (α Cygni)	1.26 summer	white	60,000	1,600
Beta Crucis	1.28 spring*	white	5,800	490
Regulus (α Leonis)	1.36 spring	blue-white	160	85
Adhara (ε Canis Majoris)	1.49 winter	white	9,000	680
Castor (α Geminorum)	1.59 winter	white	36 (12 and 12 [A] 6 and 6 [B])	45
Shaula (λ Scorpii)	1.62 summer	white	1,700	310
Bellatrix (γ Orionis)	1.64 winter	white	4,000	470
El Nath (β Tauri)	1.65 winter	white	1,700	300

Appendix 3
Greek Alphabet

The lowercase letters of the Greek alphabet are used to designate the various stars of a constellation—usually in rough order of brightness within the constellation (so that Alpha Centauri is the brightest star in the constellation Centaurus), but not always.

α	alpha	ν	nu
β	beta	ξ	xi
γ	gamma	ο	omicron
δ	delta	π	pi
ϵ	epsilon	ρ	rho
ζ	zeta	σ	sigma
η	eta	τ	tau
θ	theta	υ	upsilon
ι	iota	φ	phi
κ	kappa	χ	chi
λ	lambda	ψ	psi
μ	mu	ω	omega

Appendix 4

Giant List of Naked-Eye Magnitude Objects

The definitions of both apparent and absolute magnitude are given in the text. Unless otherwise noted, a magnitude listed is an apparent magnitude.

The few objects on this list brighter than magnitude −19 would be excruciating and no doubt soon damaging to the eye which looked at them directly. All the other objects whose apparent magnitudes are given can be seen with the naked eye under the proper conditions, with the almost definite exception of the Helix Nebula: its light is spread over an area too large to be detected with unaided vision. In addition, there are a few objects on the list which have not yet actually been observed with the naked eye, to the best of my knowledge: the asteroids Pallas and Juno, because they are very rarely as bright as the maximum values given here; Neptune, because it is so close to the limit that practically perfect conditions (in all respects) are required to make it observable to the unaided vision (for many years to come Neptune will be passing through the rich Sagittarius Milky Way and hence surely be impossible to see).

This list is more extensive and varied than any I have ever come across. A few of the more unusual entries are rather rough estimates; these are marked with two asterisks. Values which may be slightly imprecise are indicated by a single asterisk.

The figures called "light units" are based on a brightness of a magnitude of 6.0 equaling 1 unit, and indicate the relative brightness of various magnitudes.

magnitude −29.0	100 trillion (10^{14}) light units	
brightest fireballs		>−27
sun		−26.7
magnitude −24.0	1 trillion (10^{12}) light units	
absolute magnitude of M31, Great Andromeda Galaxy		−20.3*
magnitude −19.0	10 billion (10^{10}) light units	
pain threshold		−19*
absolute magnitude of a bright supernova (S Andromedae in M31, 1885)		−18.2*
very bright mock suns		>−18**
absolute magnitude of Large Magellanic Cloud		−18*
"sun-grazer" comet at perihelion (Great Sept. Comet of 1882)		−17*
absolute magnitude of Small Magellanic Cloud		−17*
full Earth as seen from moon (varies acc. to Earth cloud cover)		−16.5 to −17.5

very small diamond in "diamond ring" of total solar eclipse	−16**
magnitude −14.0 — 100 million (10^8) light units	
brightest supernova in each period of 200,000 years	−14*
star of faintest luminosity known (Van Biesbroeck's Star) if put in place of our sun	−13*
sun's corona (average)	−13*
brightest full moon, A.D. 1750–2125 (Jan. 4, 1912)	−12.98
average full moon	−12.74
Vela supernova (possibly around 4000 B.C., maybe several thousand years earlier)	−12.7**
first quarter (half-lit) moon	−9.5
Lupus supernova (A.D. 1006)	−9.5*
absolute magnitude of S Doradus, most luminous star known (if it is a single star), average (max. is twice as bright)	−9.4*
absolute magnitude of very bright nova (Nova Aquilae, 1918)	−9.3*
magnitude −9.0 — 1 million (10^6) light units	
absolute magnitude of T Coronae Borealis, "the Blaze Star," at max.	−8.4*
total brightness of night sky	−7.5**
absolute magnitude of Deneb	−7.1*
crescent moon about 30° elongation from sun	−6.5*
total magnitude of a bright section of a major aurora	>-6.0
Crab supernova (A.D. 1054)	−4 to −5
faintest fireballs (by definition)	>-4.6
Venus at best greatest brilliancy (new formula)[1]	−4.6
Venus at best greatest brilliancy (old formula)[1]	−4.4
Tycho's Star (supernova of A.D. 1572)	−4.3 ± 0.3
magnitude −4.0 — 10,000 (10^4) light units	
Halley's Comet at brightest ever recorded (April 11, A.D. 837)	−3.5*
moon during rather bright total lunar eclipse (Danjon number = 3.0)	−3.0**
Jupiter at best opposition (new formula)[1]	−3.0*
Mars at extremely excellent perihelic opposition (new formula)[1]	−2.8*
Jupiter at mean opposition (new formula)[1]	−2.55*

Brightest auroras, per discernible unit area	>−2.1**
limiting naked-eye magnitude for a person with average eyesight when sun is more than 15° alt.	−2.1*
Mars at mean opposition (new formula)[1]	−2.0*
Mercury at brightest (old formula)[1]	−1.9*
Sirius, brightest star	−1.42
Brightest nova (Nova Aquilae on June 9, 1918)	−1.4
Eta Carinae, nova-like irregular variable (maximum in April 1843)	−0.8
Canopus, second brightest star	−0.72
Saturn at best opposition (new formula)[1]	−0.4
Betelgeuse, brightest ever (December 1852)	−0.1*
Arcturus, brightest star in north celestial hemisphere	−0.06
moon during rather dark total lunar eclipse (Danjon = 2.0)	0.0*
Betelgeuse at average brightness (brightest red star)	0.7
Saturn at mean opposition (new formula)[1]	0.75
Hyades star cluster, integrated magnitude	0.8
magnitude 1.0 100 (10^2) light units	
Spica, the literally first-magnitude star	1.0
Large Magellanic Cloud (total brightness)	1.0*
Mira, most famous long-period variable, at its brightest (1779)	<1.0*
Pleiades star cluster, integrated magnitude	1.4
T Coronae Borealis, "the Blaze Star", at maxima (1866 and 1946)	2.0*
Polaris, the North Star (average)	2.0
Mars at faintest (old formula)[1]	2.0
Algol, brightest eclipsing binary, at maximum	2.15
average Perseid meteor	2.3
Mira, average maximum	3.3
Algol at minimum	3.4
Omega Centauri, brightest globular cluster	3.7
moon during extremely dark total lunar eclipse of Dec. 30, 1963 (Danjon = 0.2!)	4.1
Ganymede, brightest satellite of another planet in apparent magnitude, at a mean opposition (new formula)[1]	4.5

limiting naked-eye magnitude on clear night in large city or full-moon night in the country		4.5**
apparent magnitude of M31, Great Andromeda Galaxy		4.8
absolute magnitude of the sun		4.83
Uranus at brightest (new formula)[1]		5.3*
maximum apparent magnitude of S Andromedae, very bright supernova in M31 (1885)		5.4*
Vesta, brightest asteroid, at best oppositions		5.5*
limiting naked-eye magnitude on clear night in a medium-sized city		5.5**
M13, brightest globular cluster high in sky at 40°N latitude		5.7*
magnitude 6.0	1 (10^0) light unit	
Asteroid Ceres at best oppositions		6.4*
Helix Nebula, brightest planetary nebula in total magnitude		6.5
asteroid Pallas at rare best opposition		6.6*
limiting naked-eye magnitude on clear night in country		6.6*
asteroid Juno at very rare best opposition		7.1*
Neptune at brightest		7.6*
limiting naked-eye magnitude on the very best nights in country locations		7.5–8.0*
limiting naked-eye magnitude on best nights with special aids (observatory experiment)		8.6

NOTE: For comparison with naked-eye magnitudes, consider that very small telescopes can reach about magnitude 11.0 (0.01 or 10^{-2} light units); largest amateur telescopes about magnitude 16.0 (0.0001 or 10^{-4} light units); largest professional telescope in the world today about magnitude 21.0 (0.000001 or 10^{-6} light units) visually (better photographically); faintest object yet recorded (by the year 1977), using special image-intensifying techniques, about magnitude 25.6 (Vela pulsar, survivor of the blazing Vela supernova mentioned above)—which is not much more than 0.00000001 or 10^{-8} light units. And when the Space Telescope is put in orbit in the mid-1980s it will be able to record objects a number of times fainter still—perhaps only one hundred-millionth as bright as the dimmest object detectable to the naked eye! This should give us great respect for the abilities of man-made instruments, but consider also the fact that the unaided eye can glimpse light sources as little as one hundred-billionth as bright as the brightest it can endure.

[1] New formulas and old formulas. In the equations which predict how bright a planet will appear when seen at various distances and angles from the sun and Earth, there are some empirical values—that is, numbers which are derived from actual observation of the particular planet and the unique behavior of its brightness due to its unique complex of various properties (for example, a planet with clouds can have considerable changes in its reflectivity according to the extent, distribution, and nature of its cloud cover, but

this cannot be a factor for a planet without an atmosphere). The old formulas for planetary magnitudes are very old indeed (some at least, I believe, derive from the nineteenth century), yet they are still regularly used even in so official and widely used an observer's guide as the annual *Astronomical Almanac* (formerly *American Ephemeris and Nautical Almanac*) put out by the U.S. Naval Observatory. The "new" formulas (themselves now decades old!) are based on improved values for the empirical figures, derived from further observational studies. Some of these figures should be quite dependable, but even in the best cases it still remains true that the magnitudes of the planets are subject to at least slight variations due to physical changes on the planets and the inconstancy of the sun's light output. Mars, with its occasional planet-wide dust storms, has varied at least as much as 0.1 magnitude (the very slightest that can be detected in a good situation by an experienced variable-star observer), but Jupiter can apparently vary considerably more because of changes in its extremely active atmosphere. Uranus turns out to be considerably brighter than the old formula gave it credit for being, and so easier to see with the naked eye (outside of cities) than has been widely supposed. And Venus, which the old formula said could reach −4.4 at its most blazing instances of "greatest brilliancy," actually achieves (Steve Albers informs me) 0.18 magnitude brighter at every greatest brilliancy; on December 16, 1981, when it was (ever so marginally) brighter than it will be again for many decades, the old formula predicted it shined at −4.38, but the new formula predicted −4.56. So Venus can get even brighter than has been generally thought!

APPENDIX 5

Observing Data for Naked-Eye Planets

	Color	Mean Synodic Period (days)	Extremes of Brightness, A.D. 1950–2000[1]		Planetary Absolute Mag.[2]
			Bright	Faint	
Mercury	white[3]	115.88	−1.92	3.58	−0.36
Venus	yellow-white	583.92	−4.38	−2.57	−4.29
Earth	blue	—	—	—	−3.87
Moon	yellow-white	29.53	—	—	0.21
Mars	deep orange	779.94	−2.64	2.05	−1.52
Jupiter	yellow-white	398.88	−2.48	−1.20	−9.25
Saturn	deep gold	378.09	−0.29	1.45	−8.88 (globe)
Uranus	blue-green (in telescope)	369.66	—	—	−7.19

[1] Calculated by the computer program of Steve Albers; magnitudes are based on the "old" formulas.
[2] This is the magnitude the planet would appear to have if it were 1 Astronomical Unit from the sun and seen at full phase—that is, what it would look like if it were placed in the position of the Earth and viewed by us from the sun.
[3] I have sometimes seen Mercury take on a beautiful orange hue even to the naked eye, but this is presumably due to seeing the planet low in the sky (and hence shining through a greater thickness of air like the orange sun).

Appendix 6

Best Naked-Eye Conjunctions of Planets and Stars, 1983–1993

This selection of the "best" conjunctions is necessarily to some extent a matter of personal taste, but all the conjunctions listed are visible (and striking or interesting) to the naked eye under good conditions, except for a few conjunctions which may require binoculars or telescopes but were included because they are especially interesting. These predictions are derived from a remarkable computer program written by Steve Albers and first successfully run by him in 1977. The magnitude figures for the planets are based on the old formulas (see footnote at end of Appendix 4). The date and time may be considered universal time—which is 24-hour time based on the meridian of Greenwich, England, and must be converted to the observer's local time: for a person on eastern standard time (in the eastern U.S.), 5 hours must be subtracted (and therefore one additional hour must be subtracted for each time zone to the west). A conjunction which took place at 8 hrs. UT (universal time) took place at 3 A.M. ($8 - 5 = 3$) eastern standard time. A conjunction occurring at 0 hr. UT on March 10 (for example) would, in eastern standard time, have occurred 5 hours before the end of the previous day—7 P.M. on March 9.

The figures given are for the time of conjunction in the technical sense of "conjunction"—the time when one of the objects is due north or due south of the other. As explained in the text, this is not always precisely the closest approach. A user of this table should also be aware that if the precise time occurs when the objects are below the horizon or it is daytime at his location, the objects might not be so close together when they rise or darkness falls. In general, however, it is true that only the conjunctions involving Mercury and Venus (fast planets) are likely to be distinctly poorer just a few hours before or after the time listed. In many cases, the pairing of objects will be quite striking even for days around the actual time of conjunction (this is particularly true of conjunctions involving Jupiter and Saturn, the slower planets). Figures are computed precisely for a point near New York City, but this is of an importance of only a few seconds of arc.

This listing should include all of the most spectacular conjunctions involving planets or a planet and a star, but on many other nights the moon may also make a beautiful scene by being near a planet. There will additionally be many nights when several planets and bright stars, though not technically in conjunction, will appear in the same section of the sky, perhaps forming a striking pattern. For just a few of the endless possibilities of planet (and star) gatherings in different landscapes and weathers, see the end of Chapter Seven.

Date and time are in universal time, given as date and hour; the two celestial objects are given; S is separation in degrees and minutes of arc (and seconds of arc in a few cases), with + indicating that the first object passes north of the second, − indicating the first object passes south; E is the elongation of the conjunction from the sun in degrees east (E) or west (W) of the sun; M_1 and M_2 are the magnitudes of the first and second objects. Aqr = Aquarii; Gem = Geminorum; Leo = Leonis; Lib = Librae; Oph = Ophiuchi; Psc = Piscium; Sco = Scorpii; Sgr = Sagittarii; Tau = Tauri.

1983				S	E	M_1	M_2
Jan. 10	12	Jupiter	$Beta_1$ Sco	−0 11	W47	−1.4	2.9
Feb. 18	22	Venus	Mars	−0 32	E26	−3.4	1.4
Jun. 22	22	Jupiter	$Beta_1$ Sco	−0 10	E152	−2.1	2.9
Jul. 9	23	Venus	Regulus	−0 39	E43	−4.1	1.3
Sep. 3	23	Jupiter	$Beta_1$ Sco	−0 26	E82	−1.6	2.9
Sep. 28	21	Mars	Regulus	+0 52	W36	1.9	1.3
Dec. 17	11	Venus	Saturn	+0 9	W43	−3.7	0.8
1984							
Jan. 20	15	Jupiter	M20 (Trifid Neb.)	−0 7	W30	−1.4	—
Jan. 25	23	Venus	Neptune	+0 1 31	W35	−3.5	7.8
Jan. 27	2	Venus	Jupiter	+0 51	W35	−3.5	−1.4
Feb. 15	13	Mars	Saturn	−0 48	W100	0.6	0.7
May 17	11	Jupiter	Nu_2 Sgr	−0 1	W134	−2.1	5.0
May 21	3	Jupiter	Nu_1 Sgr	+0 2	W138	−2.1	5.0
Sep. 3	2	Mars	Antares	+2 15	E88	0.2	1.2
Oct. 13	23	Mars	Jupiter	−1 55	E75	0.5	−1.7
Nov. 19	1	Venus	Lambda Sgr	+0 0 10	E39	−3.5	2.9
Nov. 24	22	Venus	Jupiter	−2 3	E40	−3.6	−1.5
1985							
Jul. 13	23	Venus	Epsilon Tau	−0 1	W43	−3.7	3.6
Sep. 4	21	Mercury	Mars	−0 0 50	W16	−0.8	2.0
Sep. 6	10	Mercury	Regulus	+0 57	W15	−0.9	1.3
Sep. 9	1	Mars	Regulus	+0 46	W17	2.0	1.3
Sep. 21	16	Venus	Regulus	−0 26	W29	−3.4	1.3
Oct. 2	18	Mars	Chi Leo	−0 9	W25	2.0	4.7
Oct. 3	20	Venus	Chi Leo	−0 4	W26	−3.4	4.7
Dec. 16	17	Mercury	Saturn	+0 28	W21	−0.2	0.7
1986							
Feb. 7	10	Mars	$Beta_1$ Sco	−0 16	W75	1.2	2.9
Feb. 17	23	Mars	Saturn	−1 16	W80	1.1	0.7
Jun. 11	2	Mercury	Epsilon Gem	+0 0 22	E20	−0.2	3.2
Jun. 21	18	Venus	M44 (Beehive)	+0 40	E37	−3.5	3.7
Jul. 11	1	Venus	Regulus	+1 7	E41	−3.6	1.3
Aug. 31	14	Venus	Spica	−0 31	E41	−3.6	1.0
Sep. 18	2	Saturn	Nu_1 Sco	+0 2	E70	0.8	4.3
Oct. 22	16	Saturn	Psi Oph	+0 1	E38	0.8	4.6
Dec. 19	7	Mars	Jupiter	+0 31	E79	0.6	−1.9

					S	E	M_1	M_2
				1987				
Jan.	19	23	Venus	Ceres	−0 2	W47	−4.0	8.2
Jan.	24	20	Venus	Saturn	+1 47	W47	−4.0	0.8
Mar.	20	6	Ceres	M8 (Lagoon Neb.)	−0 0 20	W88	7.7	—
Apr.	19	12	Mercury	Jupiter	−1 21	W18	−0.4	−1.6
May	4	22	Venus	Jupiter	−0 38	W29	−3.3	−1.6
Nov.	20	16	Venus	Saturn	−2 8	E23	−3.3	0.7
				1988				
Feb.	22	21	Mars	Uranus	+0 1	W63	1.3	6.0
Feb.	23	13	Mars	Saturn	−1 18	W63	1.3	0.8
Mar.	6	20	Venus	Jupiter	+2 24	E44	−3.8	−1.7
Apr.	30	7	Venus	Beta Tau	−1 0	E42	−4.2	1.8
May	11	19	Venus	136 Tau	+0 0 27	E37	−4.2	4.5
Jun.	27	6	Saturn	Uranus	+1 21	E173	0.2	5.9
Oct.	4	8	Venus	Regulus	−0 14	W42	−3.6	1.3
Nov.	21	15	Saturn	M21	−0 10	E31	0.7	6.5
				1989				
Jan.	16	16	Venus	Saturn	−0 33	W19	−3.3	0.7
Mar.	12	8	Mars	Jupiter	+1 58	E69	1.3	−1.8
May	23	4	Venus	Jupiter	+0 50	E13	−3.4	−1.5
Jul.	2	4	Uranus	Vesta	−2 1	E172	5.9	5.5
Jul.	2	17	Mercury	Jupiter	−0 35	W17	−0.6	−1.5
Jul.	3	8	Saturn	28 Sgr	+0 0 0.88	E179	0.2	5.8
Jul.	11	21	Vesta	M21	+0 3	E161	5.6	6.5
Jul.	12	12	Venus	Mars	+0 28	E26	−3.3	2.0
Jul.	15	2	Uranus	SAO 186437	+0 1	E160	5.9	5.1
Jul.	15	11	Vesta	BET283	+0 4	E157	5.6	5.7
Jul.	23	11	Venus	Regulus	+1 10	E29	−3.4	1.3
Aug.	2	16	Mars	Regulus	+0 41	E19	2.0	1.3
Aug.	5	22	Mercury	Mars	+0 0 49	E18	−0.3	2.0
Aug.	19	5	Vesta	SAO 185928	−0 0 4	E123	6.1	6.1
Sep.	5	21	Uranus	Vesta	+1 55	E108	6.0	6.4
Sep.	30	14	Venus	Iota Lib	+0 0 49	E43	−3.7	4.7
Oct.	17	1	Venus	Antares	+1 48	E46	−3.8	1.2
				1990				
Feb.	3	15	Mercury	Saturn	+0 13	W25	0.1	0.8
Feb.	28	17	Mars	Saturn	−1 0	W48	1.4	0.8
Apr.	28	4	Mars	58 Aqr	−0 0 3.1	W62	1.0	6.4
Jul.	29	6	Mercury	Regulus	+0 2	E24	0.1	1.3
Aug.	12	23	Venus	Jupiter	+0 2	W21	−3.3	−1.4
Sep.	6	21	Venus	Regulus	+0 45	W15	−3.4	1.3

					S	E	M_1	M_2
				1991				
Feb.	7	7	Jupiter	M44 (Beehive)	−0 32	E169	−2.1	3.7
May	20	23	Jupiter	M44 (Beehive)	−0 34	E68	−1.5	3.7
Jun.	7	18	Mars	M44 (Beehive)	+0 7	E51	1.8	3.7
Jun.	14	5	Mars	Jupiter	+0 38	E48	1.9	−1.4
Jun.	17	23	Venus	Jupiter	+1 14	E45	−4.0	−1.4
Jun.	23	11	Venus	Mars	+0 16	E45	−4.0	1.9
Jul.	11	7	Venus	Regulus	−1 0	E42	−4.2	1.3
Jul.	14	16	Mars	Regulus	+0 41	E38	2.0	1.3
Jul.	15	8	Mercury	Jupiter	−0 5	E25	0.3	−1.3
Sep.	10	8	Jupiter	Regulus	+0 21	W18	−1.3	1.3
Sep.	10	10	Mercury	Regulus	+0 17	W18	−0.4	1.3
Sep.	10	10	Mercury	Jupiter	−0 4	W18	−0.4	−1.3
Oct.	17	3	Venus	Jupiter	−2 29	W46	−4.2	−1.4
Dec.	27	10	Venus	Zeta Lib	−0 0 10	W40	−3.6	5.6
				1992				
Jan.	10	20	Mercury	Mars	+0 39	W19	−0.2	1.6
Feb.	19	22	Venus	Mars	+0 51	W29	−3.4	1.5
Feb.	29	20	Venus	Saturn	+0 8	W27	−3.3	0.9
Mar.	6	13	Mars	Saturn	−0 26	W33	1.4	0.9
Apr.	30	18	Mars	20 Psc	0 0 1.3	W45	1.3	5.6
Aug.	23	3	Venus	Jupiter	+0 17	E19	−3.3	−1.2
Nov.	18	17	Venus	Lambda Sgr	−0 2	E40	−3.5	2.9
Dec.	21	16	Venus	Saturn	−1 4	E45	−3.8	0.9
				1993				
May	12	10	Mars	M44 (Beehive)	+0 29	E75	1.4	3.7
Jun.	22	10	Mars	Regulus	+0 46	E58	1.7	1.3
Jul.	13	17	Venus	Epsilon Tau	+0 2	E43	−3.7	3.6
Jul.	28	7	Venus	Zeta Tau (near Crab)	+0 15	W41	−3.6	3.0
Sep.	7	0	Mars	Jupiter	−0 54	E32	1.8	−1.3
Nov.	8	17	Venus	Jupiter	+0 23	W17	−3.4	1.2
Nov.	14	13	Mercury	Venus	+0 45	W16	0.6	−3.4

APPENDIX 7

Some Special Phenomena of Meteorological Optics

Phenomenon	Mechanism (in addition to reflection)	Medium	Colors	Distance from Sun (or from Antisolar Point*)
rainbow				
primary	refraction	raindrops	violet to red (working outward)	*42° (red)
secondary	refraction	raindrops	red to violet (working outward)	*51° (red)
supernumerary arcs	diffraction	raindrops	alternate pink and blue or green (as many as five or more)	*inside primary; rarely outside secondary
fog-bow (or cloud-bow or mist-bow; if full circle, sometimes called Ulloa's Ring)	refraction, diffraction	cloud droplets	white or blue, white, and orange (working outward)	*<42°

APPENDIX 7—*Continued*

Phenomenon	Mechanism (in addition to reflection)	Medium	Colors	Distance from Sun (or from Antisolar Point*)
dew-bow	refraction	dewdrops	violet to red (working toward outside of hyperbola)	*hyperbola about antisolar pt. (as usually observed)
corona and iridescent clouds	diffraction	cloud droplets or ice needles	green or blue and pink (as many as five sets in corona)	<1°–>13° (perhaps >60° rarely for irid. clouds)
Bishop's Ring (extremely rare)	diffraction	volcanic ash	blue-white inside, reddish brown outside	variable: red at 15°–28°?
glory	diffraction	cloud droplets	green or blue and pink (several sets beyond a dark or bright center)	*0°–>10°?
halo phenomena (all occur in various ice crystals) (refraction and reflection)				
small halo			white or red, white, and blue (working outward)	22°

large halo (fairly rare)	as for small halo	46°
parhelia (mock suns or sun dogs)	white or (often) red to yellow or white (sometimes other hues) to blue, working away from sun; sometimes bluish white tail outermost	≥22° to either side of sun, increasing with solar elevation up to 61° elevation (when parhelia disappear)
circumscribed halo	as for parhelia (with occasional greens and oranges!)	≥22°
upper tangential arc of circumscribed halo	as for circumscribed halo	≥22°
circumzenithal arc	red (through full range of spectrum sometimes) to violet, working inward	≥46° above sun when sun ≤32° above horizon
circumhorizontal arc (fairly rare)	as for circumzenithal arc, working outward	≥46° below sun when sun ≥58° above horizon
(reflection only)		
pillars (sun, moon, and other)	colorless themselves, take on color of light source (sun pillars often orange)	extending up (and, less often, down) from the low (or already-set) sun for a distance of >1° to >20°
parhelic (or horizontal) circle	colorless	horizontally around sky at elevation of the sun

APPENDIX 8

Solar and Lunar Eclipses

Solar Eclipses Through 1985

Date	Type	Max. Duration	Path
1982 Jan. 25	partial	—	Antarctic, New Zealand
1982 Jun. 21	partial	—	Extreme S. Africa
1982 Jul. 20	partial	—	Arctic, N.W. Europe
1982 Dec. 15	partial	—	Europe, N.E. Africa, W. Asia
1983 Jun. 11	total	5 mins., 11 secs.	Indian Ocean, Indonesia, New Guinea, Pacific Ocean
1983 Dec. 4	annular	—	Atlantic Ocean, Africa
1984 May 30	annular	—	Pacific Ocean, Mexico, S.E. United States, N. Africa
1984 Nov. 22–23	total	1 min., 59 secs.	East Indies, New Guinea, S. Pacific
1985 May 19	partial	—	Arctic
1985 Nov. 12	total	1 min., 55 secs.	Antarctic

Widely Observable Total and Annular Solar Eclipses for North America Through 2025

Date	Type	Path	Comments
1984 May 30	annular	Mexico, S.E. U.S.	sun 99.6% eclipsed (so corona should be detectable with aid); roughly 4-mile-wide path; annularity lasts 11 seconds!
1991 Jul. 11	total	Hawaii, Mexico (incl. Mexico City), Central America	6 mins., 45 secs.—longest totality for North America for more than a century; sun nearly overhead in Mexico; excellent clear-sky prospects
1992 Jan. 5	annular	part of S. California	visible at sunset in Los Angeles and San Diego
1994 May 10	annular	Baja Calif. and Arizona through Midwest to Maine and Nova Scotia	
2012 May 20	annular	Oregon, N. Calif. through Grand Canyon to central Texas	$4\frac{1}{2}$ minutes of annularity
2017 Aug. 21	total	Oregon through Midwest to South Carolina	$2\frac{1}{2}$ minutes of totality
2021 Jun. 20	annular	North of Lake Superior to Hudson Bay and north	$3\frac{1}{2}$ minutes of annularity

2023 Oct. 14	annular	Oregon through Utah to Texas, Yucatan, Venezuela, and Central America	
2024 Apr. 8	total	Mexico and Texas through St. Louis, Cincinnati, and Buffalo to Maine and Maritime Provinces.	$4\frac{1}{2}$ minutes of totality

NOTE: For more information on these North American solar eclipses, see the article by Fred Espenak in the November 1980 issue of *Astronomy* magazine. The ultimate source for basic data on solar eclipses for hundreds of years past and future used to be a work by Oppolzer, but this has been supplanted by the new *Canon of Solar Eclipses*—one of whose authors is Jean Meeus, a great calculator who is not a professional astronomer and therefore apparently does this work purely out of enthusiasm.

Partial and Total Lunar Eclipses Through 1986

* visible from United States   ** partly visible from United States

Date	Type	Dur. of Totality (mins.)	Sublunar Point[1] Long.	Lat.
1982 Jan. 9	total	84	63E	22N
1982 Jul. 6*	total	106	11W	23S
1982 Dec. 30*	total	66	171W	23N
1983 Jun. 25*	partial	—	126W	23S
1985 May 4	total	70	60W	16S
1985 Oct. 28	total	42	90E	13N
1986 Apr. 24**	total	68	168E	13S
1986 Oct. 17	total	74	67E	10N

[1] The point on Earth at which the moon appears overhead (given for time of mideclipse); the eclipse will be visible from the entire hemisphere centered on this point.

NOTE: The eclipse of July 6, 1982 features almost the longest totality possible. According to Jean Meeus, 1982 is the last calendar year for more than 500 years to contain three total eclipses of the moon. According to selenologist Patrick Moore, the remaining total lunar eclipses of this century occur on the following dates: February 20 and August 17, 1989; February 9, 1990; December 10, 1992; June 4 and November 29, 1993; April 4 and September 27, 1996; September 16, 1997; and January 21 and July 16, 2000.

NOTE: On tours to total eclipses of the sun: one organization which I have found to conduct excellent tours to totality is Gall Publications of Toronto. To the eclipse of June 11, 1983 (over 5 minutes of totality!), Gall Publications will offer tours led by a great professional astronomer, Bart Bok, and a great amateur astronomer, Jack Newton (veteran and photographer of many total solar eclipses; Newton obtained one of the first really good photographs of eclipse shadow bands ever taken). For information on these tours and possible future ones write to Gall Publications, 1293 Gerrard Street East, Toronto, Ont., Canada, M4L 1Y8, or call (416) 469-4171. Another tour to the 1983 eclipse which features an impressive array of expert guides is Sciencefaction Expeditions (3 E. 54 St., Dept. S., New York, N.Y. 10022; toll-free 1-800-223-6626, in New York [212] 751-3250).

AN ANNOTATED BIBLIOGRAPHY

An Annotated Bibliography

Works which should be particularly useful to readers of this book:

ALLEN, RICHARD HINCKLEY. *Star Names, Their Lore and Meaning*. Reprint, Dover Publications. The classic (1899) and certainly greatest work on the subject; in addition to thousands of names and tales, there are many anecdotes from the history of astronomy and even much observational information which is still useful.

BEATTY, J. KELLY; BRIAN O'LEARY; AND ANDREW CHAIKIN, editors. *The New Solar System*. Sky Publishing and Cambridge University Press. Wide-ranging collection of articles by the experts, extending through the mission of Voyager 1 at Saturn (new second edition includes results through Voyager 2 at Saturn). There is as yet no truly comprehensive work for laymen on the results of the first great period of planetary exploration; much information can be found in recent years' issues of the astronomy magazines listed below.

BURNHAM, ROBERT, JR. *Burnham's Celestial Handbook*, 3 vols. Dover Publications. A virtually complete observer's guide to the stars and other "deep sky" objects beyond the solar system for viewers using anything from the naked eye to a 10" telescope. This is a truly incomparable gathering of physical data, observational descriptions, countless photographs with a variety of instruments, and even a lot of lore (some not found in Allen)—all presented in lucid, imaginative writing by a professional astronomer who is also an amateur astronomer in the very highest sense of that worthy title.

Comets. Available from *Scientific American* magazine. Basically nontechnical articles selected from many years of *Scientific American*. Includes a collection of the magazine's original coverage of the 1910 visit of Halley's Comet.

GREENLER, ROBERT. *Rainbows, Halos, and Glories.* Cambridge University Press. Very lucid, especially thorough on halo phenomena (Greenler is one of the world's leading authorities), replete with extremely valuable and original tables, diagrams, and computer simulations. Expensive ($24.95 at this writing), but easily worth the price for the dozens of exquisite color photographs, a true connoisseur's collection. And like Ottewell, Minnaert, and Burnham, this is a writer with a truly human voice and a deep love for his subject. (The book also deals excellently with green flash, blue sky, and many other phenomena of meteorological optics.)

JOBES, GERTRUDE AND JAMES. *Outer Space: Myths, Name Meanings, Calendars.* The Scarecrow Press. This book, which seems to have gone unnoticed by astronomy writers, contains a vast amount of lore which even R. H. Allen lacks. Errors (typographical and other) are unfortunately very common, and the authors are deficient in their knowledge of astronomy. But Gertrude Jobes has a marvelous gift for storytelling, and even the most well-read student of lore will find many a new old tale in these fascinating pages.

LEY, WILLY. *Watchers of the Skies.* The Viking Press. A history of astronomy which focuses especially on the history of observation of each planet (there is even a long and interesting chapter on the asteroids). This book is not only authoritative and well documented, it is remarkably detailed, and many of those details are from the personal lives of the astronomers, so that the book takes on a human dimension which is rarely found in works of this type. The human details are not extraneous or irrelevant, but quite the opposite: they are precisely those which give us insight into the processes of scientific discovery (and often into the extent and limits of our understanding).

MINNAERT, M. *The Nature of Light and Color in the Open Air.* Translated by H. M. Kremer-Priest, revised by K. E. Brian Jay. Reprint, Dover Publications. One of the most inspiring books of any kind I have ever read. It is the classic introduction to rainbows and halos, but it is also filled with descriptions of literally hundreds of other phenomena and how to find them, and many of these phenomena are to be found treated in no other book (some of these sights Minnaert himself was the first to recognize or give any real thought to). Just one of the book's great highlights is Minnaert's exquisite account and discussion of the twilight phenomena. This book is of practical value to both scientist and artist, but is written in clear (and often lyric) language for anyone who loves the sky and nature. Minnaert was one of the world's greatest experts on meteorological optics, but the most important thing about this book is that his vast array of observations and facts embodies and conveys a spirit of extraordinary delight and wonder in the natural world.

OTTEWELL, GUY. *The Astronomical Companion.* Published by Guy Ottewell, c/o Department of Physics, Furman University, Greenville, S.C. 29613. A splendid guide to visualizing—almost to feeling as if this were a (mobile?) sculpture—the universe from full Earth and celestial sphere to full cosmos. Seventy-three giant (atlas-sized) pages in soft covei with unique color cover-painting diagrams. Guy Ottewell is an artist and writer; a scholar of language and mythology; an expert naturalist—his knowledge ranging from plants to planets, and beyond. He combines his breadth of learning, his capacity for original thought, and his talent for visualization in this series of original diagrams and many, many thousands of words of concise text that picture the universe in an ever-expanding flight of lines (graphic and verbal). There are sections on time and time-units, stellar evolution, precession, Christmas and the Star of Bethlehem, star names and their pronunciation and meaning, and virtually every major astronomical phenomenon discussed in *Wonders of the Sky*. An invaluable and beautiful reference work.

WALKER, JEARL. *The Flying Circus of Physics, with Answers.* John Wiley & Sons, Inc. Clever and concise explanations for many phenomena of meterological optics, plus a vast and varied assortment of other phenomena, from the most exotic to the most mundane (the index lists under "T" everything from tornados to toilet paper, tailgating to triboluminescence!). Contains an immense, extremely valuable bibliography (there are, for instance, nine entries even for something as rarely investigated as noctilucent clouds).

Textbooks:

BAKER, ROBERT A. *Astronomy*. Van Nostrand Co.

MITTON, SIMON, editor-in-chief. *The Cambridge Encyclopedia of Astronomy*. Crown Publishers.

RUDAUX, LUCIEN AND G. DE VAUCOULEURS. *Larousse Encyclopedia of Astronomy*. (Slightly outdated, but excellent and with absolutely classic illustrations.)

TRICKER, R. A. R. *Introduction to Meteorological Optics*. Elsevier.

Periodicals:

Annual

OTTEWELL, GUY. *Astronomical Calendar*. Similar in length and in size to *The Astronomical Companion*, mentioned above, and available from its author at the same address. This guide to all year-dependent astronomical phenomena is replete with facts and charts of all kinds. It is just about all that an astronomical almanac could be—an exquisite depiction of the year in the heavens in which the events and objects can almost speak for and present themselves. *View from the Earth*, a simplified version for children and other beginners, is also available from Ottewell for each year.

Monthly Magazines

Astronomy. 625 East St. Paul Avenue, P.O. Box 92788, Milwaukee, Wis. 53202. Aimed particularly at the amateur astronomer; excellent attention to astrophotography (both beginning and advanced), planetary exploration, and much else; currently runs a column by the author on naked-eye astronomy and star lore.

Sky and Telescope. 49 Bay State Road, Cambridge, Mass. 02238. Deservedly high reputation for decades. Articles for a wide range of experience levels; some excellent regular departments.

Weather. James Glaisher House, Grenville Place, Bracknell, Berks, RG12 1BX, England. Fairly frequent articles on halo phenomena and other sky effects of meteorological optics. Popular level to semitechnical.

The best idea is to check all of these periodicals and textbooks, and also perhaps the other books, at your local library (if they do not have a book you want, you can probably obtain it free of charge through interlibrary loan). If you wish to buy a book but cannot find it at your local store, consult the store's copy of *Books in Print* to check current price, and ask the store to order it for you.

Here is one further special publication, which must be obtained directly from its author:

GUNTER, DR. JAY. *Tonight's Asteroids*. Available for *free*, except for the price of a standard postage stamp. You receive four issues for the price of five stamps by folding up four long self-addressed, stamped envelopes into another envelope and sending it to Dr. Gunter at 1411 North Mangum Street, Durham, N.C. 27701. The four leaves (eight pages) you receive include facts about the most observable asteroids of the next two months and good charts with which to find them. Though this almost always requires a small telescope, even the naked-eye observer can benefit from the most truly original parts of the publication: Dr. Gunter's skillfully and congenially told stories about the discoverers and the (often mythological) names of the asteroids. There are also many notes of general interest to amateur astronomers, and a particularly valuable service: special issues are sent out to alert readers to the visibility of comets. Since a comet may be a new one, just discovered and only briefly visible, this kind of swift, dependable service can be essential. *Tonight's Asteroids* seems to be derived purely from Dr. Gunter's great enthusiasm for astronomy in general and the asteroids in particular—and he persuades us with his enthusiasm (and solid research) that these storied and highly individual little worlds are well worth our attention and appreciation.

INDEX

Index

A CATALOGUE OF SELECTED DOVER BOOKS
IN ALL FIELDS OF INTEREST

DRAWINGS OF WILLIAM BLAKE, William Blake. 92 plates from Book of Job, *Divine Comedy, Paradise Lost,* visionary heads, mythological figures, Laocoon, etc. Selection, introduction, commentary by Sir Geoffrey Keynes. 178pp. 8⅛ x 11. **22303-5 Pa. $4.00**

ENGRAVINGS OF HOGARTH, William Hogarth. 101 of Hogarth's greatest works: *Rake's Progress, Harlot's Progress, Illustrations for Hudibras, Before and After, Beer Street and Gin Lane,* many more. Full commentary. 256pp. 11 x 13¾. **22479-1 Pa. $12.95**

DAUMIER: 120 GREAT LITHOGRAPHS, Honore Daumier. Wide-ranging collection of lithographs by the greatest caricaturist of the 19th century. Concentrates on eternally popular series on lawyers, on married life, on liberated women, etc. Selection, introduction, and notes on plates by Charles F. Ramus. Total of 158pp. 9⅜ x 12¼. **23512-2 Pa. $6.00**

DRAWINGS OF MUCHA, Alphonse Maria Mucha. Work reveals draftsman of highest caliber: studies for famous posters and paintings, renderings for book illustrations and ads, etc. 70 works, 9 in color; including 6 items not drawings. Introduction. List of illustrations. 72pp. 9⅜ x 12¼. (Available in U.S. only) **23672-2 Pa. $4.00**

GIOVANNI BATTISTA PIRANESI: DRAWINGS IN THE PIERPONT MORGAN LIBRARY, Giovanni Battista Piranesi. For first time ever all of Morgan Library's collection, world's largest. 167 illustrations of rare Piranesi drawings—archeological, architectural, decorative and visionary. Essay, detailed list of drawings, chronology, captions. Edited by Felice Stampfle. 144pp. 9⅜ x 12¼. **23714-1 Pa. $7.50**

NEW YORK ETCHINGS (1905-1949), John Sloan. All of important American artist's N.Y. life etchings. 67 works include some of his best art; also lively historical record—Greenwich Village, tenement scenes. Edited by Sloan's widow. Introduction and captions. 79pp. 8⅜ x 11¼. **23651-X Pa. $4.00**

CHINESE PAINTING AND CALLIGRAPHY: A PICTORIAL SURVEY, Wan-go Weng. 69 fine examples from John M. Crawford's matchless private collection: landscapes, birds, flowers, human figures, etc., plus calligraphy. Every basic form included: hanging scrolls, handscrolls, album leaves, fans, etc. 109 illustrations. Introduction. Captions. 192pp. 8⅞ x 11¾. **23707-9 Pa. $7.95**

DRAWINGS OF REMBRANDT, edited by Seymour Slive. Updated Lippmann, Hofstede de Groot edition, with definitive scholarly apparatus. All portraits, biblical sketches, landscapes, nudes, Oriental figures, classical studies, together with selection of work by followers. 550 illustrations. Total of 630pp. 9⅛ x 12¼. **21485-0, 21486-9 Pa., Two-vol. set $15.00**

THE DISASTERS OF WAR, Francisco Goya. 83 etchings record horrors of Napoleonic wars in Spain and war in general. Reprint of 1st edition, plus 3 additional plates. Introduction by Philip Hofer. 97pp. 9⅜ x 8¼. **21872-4 Pa. $4.00**

CATALOGUE OF DOVER BOOKS

THE SENSE OF BEAUTY, George Santayana. Masterfully written discussion of nature of beauty, materials of beauty, form, expression; art, literature, social sciences all involved. 168pp. 5⅜ x 8½. 20238-0 Pa. $3.00

ON THE IMPROVEMENT OF THE UNDERSTANDING, Benedict Spinoza. Also contains *Ethics, Correspondence,* all in excellent R. Elwes translation. Basic works on entry to philosophy, pantheism, exchange of ideas with great contemporaries. 402pp. 5⅜ x 8½. 20250-X Pa. $4.50

THE TRAGIC SENSE OF LIFE, Miguel de Unamuno. Acknowledged masterpiece of existential literature, one of most important books of 20th century. Introduction by Madariaga. 367pp. 5⅜ x 8½.
20257-7 Pa. $4.50

THE GUIDE FOR THE PERPLEXED, Moses Maimonides. Great classic of medieval Judaism attempts to reconcile revealed religion (Pentateuch, commentaries) with Aristotelian philosophy. Important historically, still relevant in problems. Unabridged Friedlander translation. Total of 473pp. 5⅜ x 8½. 20351-4 Pa. $6.00

THE I CHING (THE BOOK OF CHANGES), translated by James Legge. Complete translation of basic text plus appendices by Confucius, and Chinese commentary of most penetrating divination manual ever prepared. Indispensable to study of early Oriental civilizations, to modern inquiring reader. 448pp. 5⅜ x 8½. 21062-6 Pa. $5.00

THE EGYPTIAN BOOK OF THE DEAD, E. A. Wallis Budge. Complete reproduction of Ani's papyrus, finest ever found. Full hieroglyphic text, interlinear transliteration, word for word translation, smooth translation. Basic work, for Egyptology, for modern study of psychic matters. Total of 533pp. 6½ x 9¼. (Available in U.S. only) 21866-X Pa. $5.95

THE GODS OF THE EGYPTIANS, E. A. Wallis Budge. Never excelled for richness, fullness: all gods, goddesses, demons, mythical figures of Ancient Egypt; their legends, rites, incarnations, variations, powers, etc. Many hieroglyphic texts cited. Over 225 illustrations, plus 6 color plates. Total of 988pp. 6⅛ x 9¼. (Available in U.S. only)
22055-9, 22056-7 Pa., Two-vol. set $16.00

THE STANDARD BOOK OF QUILT MAKING AND COLLECTING, Marguerite Ickis. Full information, full-sized patterns for making 46 traditional quilts, also 150 other patterns. Quilted cloths, lame, satin quilts, etc. 483 illustrations. 273pp. 6⅞ x 9⅝. 20582-7 Pa. $4.95

CORAL GARDENS AND THEIR MAGIC, Bronsilaw Malinowski. Classic study of the methods of tilling the soil and of agricultural rites in the Trobriand Islands of Melanesia. Author is one of the most important figures in the field of modern social anthropology. 143 illustrations. Indexes. Total of 911pp. of text. 5⅝ x 8¼. (Available in U.S. only)
23597-1 Pa. $12.95

TONE POEMS, SERIES II: TILL EULENSPIEGELS LUSTIGE STREICHE, ALSO SPRACH ZARATHUSTRA, AND EIN HELDENLEBEN, Richard Strauss. Three important orchestral works, including very popular *Till Eulenspiegel's Marry Pranks,* reproduced in full score from original editions. Study score. 315pp. 9⅜ x 12¼. (Available in U.S. only)
23755-9 Pa. $8.95

TONE POEMS, SERIES I: DON JUAN, TOD UND VERKLARUNG AND DON QUIXOTE, Richard Strauss. Three of the most often performed and recorded works in entire orchestral repertoire, reproduced in full score from original editions. Study score. 286pp. 9⅜ x 12¼. (Available in U.S. only) 23754-0 Pa. $7.50

11 LATE STRING QUARTETS, Franz Joseph Haydn. The form which Haydn defined and "brought to perfection." *(Grove's).* 11 string quartets in complete score, his last and his best. The first in a projected series of the complete Haydn string quartets. Reliable modern Eulenberg edition, otherwise difficult to obtain. 320pp. 8⅜ x 11¼. (Available in U.S. only)
23753-2 Pa. $7.50

FOURTH, FIFTH AND SIXTH SYMPHONIES IN FULL SCORE, Peter Ilyitch Tchaikovsky. Complete orchestral scores of Symphony No. 4 in F Minor, Op. 36; Symphony No. 5 in E Minor, Op. 64; Symphony No. 6 in B Minor, "Pathetique," Op. 74. Bretikopf & Hartel eds. Study score. 480pp. 9⅜ x 12¼. 23861-X Pa. $10.95

THE MARRIAGE OF FIGARO: COMPLETE SCORE, Wolfgang A. Mozart. Finest comic opera ever written. Full score, not to be confused with piano renderings. Peters edition. Study score. 448pp. 9⅜ x 12¼. (Available in U.S. only) 23751-6 Pa. $11.95

"IMAGE" ON THE ART AND EVOLUTION OF THE FILM, edited by Marshall Deutelbaum. Pioneering book brings together for first time 38 groundbreaking articles on early silent films from *Image* and 263 illustrations newly shot from rare prints in the collection of the International Museum of Photography. A landmark work. Index. 256pp. 8¼ x 11.
23777-X Pa. $8.95

AROUND-THE-WORLD COOKY BOOK, Lois Lintner Sumption and Marguerite Lintner Ashbrook. 373 cooky and frosting recipes from 28 countries (America, Austria, China, Russia, Italy, etc.) include Viennese kisses, rice wafers, London strips, lady fingers, hony, sugar spice, maple cookies, etc. Clear instructions. All tested. 38 drawings. 182pp. 5⅜ x 8.
23802-4 Pa. $2.50

THE ART NOUVEAU STYLE, edited by Roberta Waddell. 579 rare photographs, not available elsewhere, of works in jewelry, metalwork, glass, ceramics, textiles, architecture and furniture by 175 artists—Mucha, Seguy, Lalique, Tiffany, Gaudin, Hohlwein, Saarinen, and many others. 288pp. 8⅜ x 11¼. 23515-7 Pa. $6.95

THE AMERICAN SENATOR, Anthony Trollope. Little known, long unavailable Trollope novel on a grand scale. Here are humorous comment on American vs. English culture, and stunning portrayal of a heroine/villainess. Superb evocation of Victorian village life. 561pp. 5⅜ x 8½.
23801-6 Pa. $6.00

WAS IT MURDER? James Hilton. The author of *Lost Horizon* and *Goodbye, Mr. Chips* wrote one detective novel (under a pen-name) which was quickly forgotten and virtually lost, even at the height of Hilton's fame. This edition brings it back—a finely crafted public school puzzle resplendent with Hilton's stylish atmosphere. A thoroughly English thriller by the creator of Shangri-la. 252pp. 5⅜ x 8. (Available in U.S. only)
23774-5 Pa. $3.00

CENTRAL PARK: A PHOTOGRAPHIC GUIDE, Victor Laredo and Henry Hope Reed. 121 superb photographs show dramatic views of Central Park: Bethesda Fountain, Cleopatra's Needle, Sheep Meadow, the Blockhouse, plus people engaged in many park activities: ice skating, bike riding, etc. Captions by former Curator of Central Park, Henry Hope Reed, provide historical view, changes, etc. Also photos of N.Y. landmarks on park's periphery. 96pp. 8½ x 11. 23750-8 Pa. $4.50

NANTUCKET IN THE NINETEENTH CENTURY, Clay Lancaster. 180 rare photographs, stereographs, maps, drawings and floor plans recreate unique American island society. Authentic scenes of shipwreck, lighthouses, streets, homes are arranged in geographic sequence to provide walking-tour guide to old Nantucket existing today. Introduction, captions. 160pp. 8⅞ x 11¾. 23747-8 Pa. $6.95

STONE AND MAN: A PHOTOGRAPHIC EXPLORATION, Andreas Feininger. 106 photographs by *Life* photographer Feininger portray man's deep passion for stone through the ages. Stonehenge-like megaliths, fortified towns, sculpted marble and crumbling tenements show textures, beauties, fascination. 128pp. 9¼ x 10¾. 23756-7 Pa. $5.95

CIRCLES, A MATHEMATICAL VIEW, D. Pedoe. Fundamental aspects of college geometry, non-Euclidean geometry, and other branches of mathematics: representing circle by point. Poincare model, isoperimetric property, etc. Stimulating recreational reading. 66 figures. 96pp. 5⅝ x 8¼.
63698-4 Pa. $2.75

THE DISCOVERY OF NEPTUNE, Morton Grosser. Dramatic scientific history of the investigations leading up to the actual discovery of the eighth planet of our solar system. Lucid, well-researched book by well-known historian of science. 172pp. 5⅜ x 8½. 23726-5 Pa. $3.50

THE DEVIL'S DICTIONARY. Ambrose Bierce. Barbed, bitter, brilliant witticisms in the form of a dictionary. Best, most ferocious satire America has produced. 145pp. 5⅜ x 8½. 20487-1 Pa. $2.25

THE CHICAGO
PUBLIC LIBRARY
QB
44.2
.S33
1983
C
P
L
FOR REFERENCE USE ONLY
Not to be taken from this building
Business/Science/Technology
Division